THE INTERNATIONAL
URANIUM MARKET

THE INTERNATIONAL URANIUM MARKET

Thomas L. Neff

BALLINGER PUBLISHING COMPANY
Cambridge, Massachusetts
A Subsidiary of Harper & Row, Publishers, Inc.

Research on which this work is based was supported in part by the U.S. Department of Energy which retains non-exclusive, irrevocable license to use and dispose of results of that research contained herein.

International Standard Book Number: 0-88410-850-3

Library of Congress Catalog Card Number: 84-9255

Printed in the United States of America

Library of Congress Cataloging in Publication Data

Neff, Thomas L.
   The international uranium market.

   Includes index.
   1. Uranium industry. I. Title.
HD9539.U7N44   1984   382′.424932   84-9255
ISBN 0-88410-850-3

To Sylvia—
best critic, best friend

CONTENTS

FIGURES

TABLES

PREFACE AND ACKNOWLEDGMENTS

Since the 1940s, the international availability of uranium supplies has grown increasingly important to many nations. This importance stems from two facts: that uranium may be used in the making of nuclear weapons, and that it may also be used to generate electricity in commercial nuclear power plants. The first, together with the spread of nuclear technology, gives rise to national and international security concerns associated with the potential for nuclear weapons proliferation. The second fact, together with major commitments worldwide to nuclear power as an alternative to oil, has made international supplies of uranium a central issue of energy security for many nations. These concerns have not, in fact, remained separated, with the consequence that national and international efforts to cope with the dual nature of the atom have often conflicted.

The terms of uranium trade are also of concern to producer nation governments. Some, like Canada and Australia, are sensitive to the proliferation issue; other producers may place a higher emphasis on commercial opportunities. All are sensitive to the value of their natural resources and the rents that might be taken. Consumer nations have worried about the adequacy of uranium resources, the stability of suppliers, and future uranium prices. Expectations about these matters have strongly affected national nuclear technology development programs, leading to massive expenditures on reprocessing and breeder reactor technologies.

For the United States, nuclear fuel trade has long been a locus of difficult policy choices. In the 1960s, the United States sought to protect its domestic uranium industry from foreign competition. In the next decade, it tried to use its rapidly eroding nuclear fuel supply leverage to compel the actions of other nuclear supplier nations, while simultaneously promoting assured supplies of conventional nuclear fuel as a nonproliferation strategy. In the 1980s, the United States finds its supplier role shrinking and its domestic uranium industry again endangered by low-cost imports.

Despite the commercial and energy security importance of international uranium trade—which is second only to oil among energy commodities—and despite the large number of critical policy issues at stake, there has been little clear integrative understanding of the nature of the international nuclear fuel market. In part this is because of the high level of secrecy associated with the nuclear industry, as well as the usual commercial reluctance to disclose information about investment, sales, and trade patterns. It is also due to the inherent difficulty of assessing a market where both economic and noneconomic forces play essential roles and where radical change is almost a given.

This study seeks to remedy this deficiency, to provide the basic information needed by industry, governments, and concerned citizens, and to develop an understanding of market structure and behavior that will assist both in assessing prospective market evolution and in analyzing a host of important policy issues. It also seeks, more broadly, to advance our understanding of international energy markets, where governments and relations between governments play essential roles but where economic tensions cannot long be ignored. It would be difficult to find a market more suited to the study of the interplay between economic and noneconomic forces.

The research on which this book is based was conducted over a period of nearly seven years. My interest was originally stimulted by discussions with Hans Landsberg, Spurgeon Keeny, and other participants during work on the Nuclear Energy Policy Study sponsored by the Ford Foundation in 1976–77.

This interest developed into a series of contract research efforts initiated in collaboration with Henry D. Jacoby when I came to the M.I.T. Energy Laboratory in 1977. These efforts were supported by the United States Department of Energy and included studies of nuclear fuel assurance issues, relationships between nuclear fuel markets and nuclear technology development (in collaboration with Michael Driscoll of M.I.T.'s Department of Nuclear Engineering), and U.S. roles in international uranium and enrichment markets. Research support also came from M.I.T.'s Center for Energy Policy Research, the Overseas Electrical Industry Survey Institute (Japan), and from the M.I.T./Japan Endowment for International Energy Policy Studies.

A number of M.I.T. students participated in this ongoing research program and directly or indirectly made contributions to this book. They include Richard Charpie, Tatsujiro Suzuki, Murray Kenney, Karen von Bismarck,

Isi Saragosi, Gordon Swartz, Jerry Hammond, Markus Rohrbasser, Alvin Streeter, Wendy Newman, Armando Zamora Reyes, Lily Ling, and Aiko Mondori.

I am also grateful to colleagues at M.I.T. and Harvard who challenged and helped guide research on uranium markets. Eugene Skolnikoff and Albert Carnesale organized several international nuclear policy meetings with German and Japanese participants that revealed important national differences of perspective and created opportunities for numerous discussions of nuclear markets and nuclear policies. Ted Greenwood shared his research on Canadian energy policy and U.S.–Canadian relationships; Robert Rotberg, research on South Africa and Namibia; Michael Driscoll, explorations of fuel utilization in nuclear power plants and potential for extraction of uranium from seawater; John Houghton, research on uranium geology and assessment methodologies; George Rathjens, an interest in nonproliferation policy questions and work on enrichment supply and demand; and Paul Joskow, studies of the Westinghouse case and utility decisionmaking. I also benefitted from discussions with Abram Chayes, Joseph Nye, Harvey Brooks, and Richard Lester.

A number of industry experts provided information and guidance, though not all might agree with my conclusions. Don Couchman, James Colby, Carl Walske, France Millet, and A. von Kienlin helped identify the important questions and critiqued early efforts. I am also indebted to Ralph Brittelli, Jr., Nicholas Nikazmerad, Terrance Price, James Bedore, Jeff Combs, Kazuhisa Mori, Hiroshi Murata, M. Stark, and Rayden Crawley. I am also thankful to a number of electric utility executives, including James Seery, Akira Yamada, Gordon Corey, Fujio Sakagami, Noboru Morioka, Klaus Peter Messer, Mason Willrich, and Shigefumi Tamiya.

Government policy and program officials and advisors in several nations have been both helpful and encouraging in my assessments of markets and policy environments. In the United States, I am particularly grateful to Edward Milenky, Fred McGoldrick, Gene Clark, John Patterson, Dave Thomas, Eleanor Steinberg, Roger Gagne, Audrey Flieger, Lawrence Scheinman, Warren Donnelly, Leonard Weiss, Charles Trabant, Dan Nikodem, S. Victoria Krusiewski, and Mike Telson. Clark Huffman and Betty Blevins provided essential and much appreciated information. In Australia, Denis Ives and other officials critiqued early drafts and Michael Moinard gave useful suggestions for the final version. R.M. Williams was willing to share some of his extensive knowledge of Canadian and nonCanadian energy matters and R.W. Morrison provided useful information. In Japan, Atsuhiko Yatabe and Tetsuya Endo provided insights into international policy issues and Tomihiro Taniguchi and Masaichi Kamimura have lent interest and support to nuclear fuels policy research.

I have also benefitted from discussions with—and the work of—Edward Wonder, Myron Kratzer, John Gray, Kazuya Fujime, Steven Warnecke, Tomohiro Toichi, Ernesto Villarreal, Atsuyuki Suzuki, DeVerle Harris, Joen Greenwood, Philip Farley, Kumao Kaneko, Karl Kaiser, Keiichi Oshima, Takeo Iguchi, Jeremy Platt, Julian Steyn, and others. Panos Cavoulacos and Richard Lester carefully read drafts of particular chapters. Of course, all responsibility for error remains with the author.

Several colleagues have played central roles in formulating and conducting the research on which this book is based. M.A. Adelman pioneered research on international energy markets and has shared his experience, ideas, time, and gentle criticism at various stages of this work. Henry Jacoby collaborated in the development of research on international nuclear fuel markets and in the writing of several articles based on this work. The approach taken at several junctures in this book reflects his valued conceptual contributions. Michael Lynch played a central role in the research underlying Part II of this book, developing detailed information on mines and contracting worldwide and contributing to the evaluation of the domestic policy context for participation by African nations in the international market. Chapters 5 and 6 are based, in part, on working papers for which we were coauthors.

In addition to these contributions, several dedicated and talented staff members of the Energy Laboratory have made important contributions to research progress. Virginia Faust and Gianna Sabella helped collect and organize an extensive data base. This work, transformed by technological revolution (expressed in Lotus 1–2–3), has been carried to elegant conclusion by Mary Alice Sanderson. Shelley Rosenstein contributed her great skills as a research librarian; I, at least, was never sorry.

The preparation of this book would have been impossible without assistance from two sources. The Center for Energy Policy Research at M.I.T. helped capitalize on and extend externally supported research into policy-related areas at critical points over the years. I am grateful to the contributors to the CEPR and to its director, Loren Cox, for this support. Equally important, if not more so given the difficulty of conducting a coherent ongoing research program in the funding environment of the past few years, has been the continued support and encouragement from an endowment contributed to M.I.T. for international energy policy studies by the farsighted government of Japan in 1979.

INTRODUCTION

For more than forty years, uranium has been traded in international commerce. This trade originated in the nuclear weapons procurement programs of the United States and the United Kingdom, and subsequently that of France. Early shipments from Canada and the Belgian Congo (now Zaire)— and later from South Africa, Australia, Gabon, and Niger—played essential roles in the development of nuclear weapons and the expansion of nuclear arsenals in the major weapons states. This weapons-related trade peaked in 1959, when international shipments totaled nearly 20,000 metric tonnes of uranium, enough for more than five thousand primitive nuclear explosive devices (or many more of sophisticated design).

As weapons demand saturated, international trade in uranium declined precipitously. The industry entered a depressed state, with worldwide production and prices falling nearly 60 percent by 1966 despite universal expectations of a promising new era of commercial nuclear power development. During this period, governments intervened to protect domestic mining and milling industries. When the first stages of nuclear power development came, with a wave of reactor orders in the United States, the new market was effectively isolated from foreign supply by a U.S. government prohibition on imports for domestic use. Commercial sales outside the United States grew slowly, reaching only 2,000 metric tonnes in 1968—a tenth of the weapons peak.

It was not until the fourth decade of international uranium trade in the early 1970s that a commercial market began to emerge. But its birth was far from calm. A number of events conspired to create a virtual panic, resulting in a dramatic escalation of prices and massive new commitments to future supplies of uranium. During this period a cartel was formed (itself the product of the industry depression); the United States announced the opening of its market to foreign imports and stimulated new demand for uranium worldwide by demanding new commitments to enrichment services (over which it had a near monopoly); a major private supplier contracted to supply large amounts of uranium it did not have; India tested a nuclear explosive device and, together with aggressive international marketing of sensitive nuclear technology, precipitated a crisis of confidence in the world nonproliferation regime and disruptive actions by Canada, the United States, and other nations.

All this occurred in an era of heightened concern among consumer nations about the security of energy supplies, first about oil and then about uranium for the nuclear power plants that many nations were developing to reduce their dependence on imported oil.

Today, the world nuclear fuel market has grown beyond its origins in weapons programs. In terms of energy content—the number of kilowatt-hours that might be generated—international shipments of uranium are equivalent to about five million barrels a day of oil or 400 million metric tonnes of steam coal annually (several times the actual volume of steam coal trade). The economic value of international nuclear fuel trade is second only to oil in international trade in energy commodities. The international market, however, is still subject to the complex web of economic and political forces that have so strongly affected it in the past.

While spot market prices for nuclear fuel have dropped, in constant dollar terms, to the depressed levels of the 1960s, most existing international supply arrangements reflect the panic of the 1970s. The legacy of that panic endures in the form of contractual rigidities and sovereign intervention and the ensuing importance of supply security, resource nationalism, and the international security imperative of nonproliferation.

This book is concerned with the structure and evolution of the international uranium market and the economic and noneconomic forces that shape it. In Part I, we examine the structure of the international market, its historical development, and basic conditions of supply and demand. As discussed in Chapter 1, the structure of the nuclear fuel market is technically more complex than that of most energy commodities: uranium must be mined, refined, converted in chemical form, enriched and, finally, fabricated into fuel elements for a particular nuclear power plant. These activities take place in coupled markets characterized by a rather high degree of concentration and—for enrichment at least—a significant degree of technical secrecy. Governments play essential roles in virtually all the nations participating in these markets. Largely because of the intrinsic connection to the problem of nuclear weapons proliferation, international market structure is also institutionally complex.

In Chapter 2, we trace the history of international nuclear fuel trade from its origin in military programs, through the era of conflict and instability, and into the present period of market adjustment and policy accommodation. Chapter 3 considers what is turning out to be a rapidly changing view of uranium as a natural resource, and examines trends in exploration and discovery, resource economics, and production. In Chapter 4, we consider the nature of uranium demand and its relationship to nuclear power growth, enrichment contracting, inventory desires, and technological change in the nuclear fuel cycle.

In Part II, we turn to an analysis of the uranium industry in key producer nations, and the policy environment affecting each nation's role in international nuclear fuel trade. The industry analysis is conducted on a mine-by-mine basis and includes consideration of resource characteristics, ownership and foreign participation, and export commitments. The latter are generally closely held secrets, but have been estimated here from public and private sources. It is hoped that the comprehensive view developed can compensate for minor inaccuracies in details. The roles of Canada and Australia are treated in Chapter 5 and the African producers in Chapter 6.

While the international market remains dominated by the same countries that were the source of supplies for weapons programs, there is now a growing fringe of smaller producer nations, as discussed in Chapter 7. This chapter also examines the important role of France in international nuclear fuel markets: France is simultaneously a producer, importer and exporter of uranium, as well as a supplier of reactors, enrichment, reprocessing, and other fuel cycle services. The role of the United States as an importer and exporter is also considered briefly here.

Part III integrates the analyses of Parts I and II into a comprehensive overview of international nuclear fuel trade and builds on this analysis to explore the future of this strategically important market. Chapter 8 considers supply and demand from a global perspective, focusing first on overall supply and demand balances, relationships between uranium and enrichment contracting, and nuclear fuel inventories. It then considers the specific situations of the major consumer nations, France, Japan and West Germany, which together account for more than two-thirds of international commitments to uranium and an even higher fraction of enrichment services.

In Chapter 9, we examine changing relationships between the United States and international uranium markets. The role of U.S. buyers in the future international market is the single most important factor affecting the evolution of that market over the remainder of this decade, and is second only to the fate of nuclear power itself in the years beyond. A central issue here is whether the United States will enter the world market in a significant way, or isolate itself behind protectionist measures as it did in the past. The outcome will affect world prices, investment, and security in its several dimensions.

Chapter 10 summarizes the major conclusions of the book, identifies the factors that will most significantly affect the future market, and seeks to characterize that future. A central finding is that the resource endowment internationally is much richer than thought in the 1970s and that basic economic forces would, acting alone, result in uranium prices that did not increase greatly in real terms over at least the next several decades. However, because of existing rigidities and sovereign involvements, primary market conditions will continue to deviate significantly from those that would be economically efficient over at least the remainder of this decade.

During this period, secondary market transactions will continue to redistribute the resulting gains and losses (but not reduce them greatly), and prices in these markets may more closely approximate long-run resource costs than will existing contract prices set in the panic years of the last decade. This expectation of low long-run resource costs does not guarantee that the longer term price-and-supply trajectory will be smooth: high levels of sovereign intervention and persistent instabilities make it likely that there will be discontinuities and significant fluctuations about the economically efficient path, and the resultant uncertainties will lead to higher risk premia for many investments and transactions, increasing the cost of nuclear fuel and nuclear power.

PART I
STRUCTURE AND ORIGINS

1 STRUCTURE OF THE FUEL-SUPPLY SYSTEM

The structure of the nuclear fuel system is quite different from that of any other internationally traded energy commodity, and these differences affect the price and availability of fuel internationally. This system is technically complex, with a sequence of interdependent processing steps that link conditions of uranium supply to those of enrichment and other processing steps. This technical complexity also makes it difficult for many nations to maintain a completely independent fuel-cycle capability, thus ensuring the need for international transactions even beyond those associated with uranium procurement itself.

Supplies of critical materials and services are concentrated in a few countries, leading to concerns about the dependability and security of supply. As with other strategically important commodities, both political and economic complexities are thus introduced. In the case of nuclear fuel, the central political complexity arises from the potential connections between nuclear power and the problem of nuclear weapons, and from the fabric of treaties, controls, and safeguards that have been designed to restrain the proliferation of weapons capability.

In the following sections, we first examine the technical structure of the nuclear fuel-supply system and explore the implications of this structure for uranium and nuclear fuel supply and demand. We then turn to a review of the coupled international markets for uranium and enrichment services. Finally we examine the fabric of political and institutional constraints that so strongly affect uranium and nuclear fuel trade internationally.

3

TECHNICAL STRUCTURE

While a great number of types of reactors have been developed over the past several decades, the vast majority of commercial power reactors are light-water reactors (LWRs) fueled by enriched uranium and moderated by ordinary water (as opposed to deuterium-rich heavy water), and this dominance seems likely to increase in the future. A few reactors operate on somewhat different fuel cycles. In subsequent chapters, much of our discussion focuses on the fuel needs of LWRs, though the requirements of other types of reactors are included in the analysis.

Fuel for a light-water reactor is the result of a long series of processing steps that begins with the mining of uranium-bearing ores and ends with a batch of fuel assemblies that are used to replace, approximately annually, one-fifth to one-third of the total fuel material in a reactor. In processing, the uranium ore is milled to recover the 0.05 percent to 10 percent or more uranium contained in it. The result is "yellowcake": U_3O_8 with some impurities. The yellowcake is then purified, and the uranium converted to a new chemical compound, uranium hexafluoride (UF_6).

Uranium contains only about 0.711 percent of the fissile isotope U-235, the remainder being U-238. Of these, only U-235 can be fissioned by the low-energy neutrons that mediate the chain reaction in a conventional reactor. Since the concentration of U-235 in natural uranium is too low to sustain a chain reaction in an LWR, the proportion of this isotope must be increased to about 3 percent by isotopic enrichment, a technology that has been developed commercially by only a few countries. A fraction of the original U-235—variable, within limits, by the enricher—remains as "tails" from the enrichment process. After it leaves the enrichment plant, the enriched UF_6 goes to a fuel fabricator where it is converted to uranium dioxide (UO_2), formed into pellets, and fabricated into fuel assemblies. Fuel fabricated for one reactor generally cannot be used in another.

This sequence of processing steps is more complicated than for other energy forms, and it requires more time. An idealized procurement schedule for a pressurized-water reactor, one of the two main types of LWR, is shown in Figure 1-1. Each step in the process is illustrated by a rectangle, and the first core (or full) loading and several reloads are shown. The height of each rectangle is a rough indication of the quantity of material or fuel-cycle services involved in that step, and the length represents the time required. Note that the manufacture of the first core requires more inputs than reloads. Roughly three years may be required to produce the initial fueling, and reloads take more than twenty months. When there are uncertainties—as when international purchases are involved, when there is a need to invest in mines and mills in order to ensure supply, or when renegotiation of contracts may be required—utilities may allow still more time for some

Figure 1–1.   Nuclear Fuel-Cycle Supply Logistics.

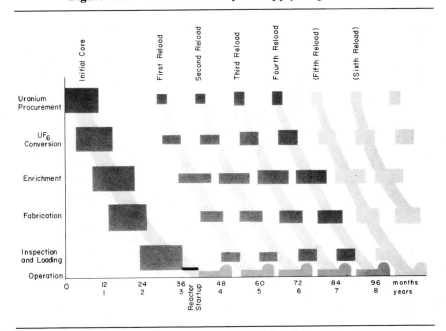

Note: Nuclear fuel-cycle lead times for a pressurized-water reactor. The shaded blocks show the flow of fuel material through the fuel cycle. The length of each block designates the time needed for each process step, while the height of the block designates the quantity of material or services involved. The first fuel loading requires more than three years to produce and considerably more materials and services than do reloads.

fuel-cycle steps. Table 1–1 shows uranium and enrichment needs for various reactor types and procurement lead times; these parameters are used subsequently in this book to calculate reactor requirements and fuel-cycle flows.

The long lead times associated with fuel supply have significant effects. Procurement of uranium, the first stage, must be undertaken well in advance of actual reactor needs. Consequently, decisions about uranium are subject to the uncertainties and long lead times that attend nuclear power generally as well as those that accompany the intermediate processing steps.

But there is a compensating virtue: interruption at early stages of the fuel cycle would not have an immediate effect on output. For example, failure of delivery from a natural uranium supplier could not result in an interruption of electric generation for at least two years. At current inventory levels of three or more years of forward supply, such disruptions might not have a physical impact for five or more years. This is very different from the situation with oil where near-term crises develop rapidly: the length of time

Table 1-1. Reactor Fuel Requirements and Lead Times.[a]

Reactor Type	LWR (no recycle)	HWR	HTR	GCR	AGR
Initial Core					
Natural uranium (MTU/GWe)	363	145	236	918	458
Separative work (MTSWU/GWe)	243	—	310	—	252
Reloads					
Natural uranium (MTU/GWe)	138	119	57	214	131
Separative work (MTSWU/GWe)	111	—	75	—	89

Fuel Cycle Lead Times

	First Core	Reloads
All but HWR		
Enrichment	2 years	Same calendar year
Natural uranium	3 years	1 year
HWR		
Natural uranium	2 years	1 year

a. Based on fuel-cycle characteristics given in Reference 2 with adjustments to 70 percent reactor capacity factor and 0.20 percent tails assay where enrichment is required.

Note: The predominant reactor type worldwide is the Light Water Reactor (LWR) of which there are two types, the Pressurized Water Reactor (PWR) and the Boiling Water Reactor (BWR). The fuel cycle requirements shown are for a mix of 2/3 PWRs and 1/3 BWRs. The Heavy Water Reactor (HWR) operates on natural uranium (unenriched) but is moderated by heavy water (deuterium). The High-Temperature Reactor (HTR) utilizes enriched uranium but is gas-cooled and graphite-moderated. Great Britain and France still operate Gas-Cooled Reactors (GCR) that utilize natural uranium. New versions of these reactors are Advanced Gas Reactors (AGR), utilizing enriched uranium. Reactor fuel needs (roughly) scale with the size of the reactor, stated in units of electrical output, or MWe (new LWRs have capacities of about 1100 MWe). Uranium requirements in the table are stated in terms of the uranium metal content—in metric tonnes uranium (MTU)—of original natural uranium feed material, prior to enrichment. Required enrichment services are stated in metric tonnes Separative Work. (MTSWU).

between supplier failure and impact on economic activity would rarely exceed three months for oil.

The natural flywheel effect of the nuclear supply system is enhanced by the conservative planning of some consumers and suppliers. Utilities often order fuel on the assumption that the reactor will operate at a 75 to 80 percent capacity factor. In practice, reactors have been operating at significantly lower average capacity factors, resulting in reduced fuel consumption rates. Table 1-2 shows average capacity factors achieved in 1982 among different nations and for various types of reactors.

Table 1-2. Average and Cumulative Load Factors for Nuclear Power Plants in Selected Nations (1982).

Country	Annual Load Factor % (average)	Number and Output of Reactors	Lifetime Cumulative Load Factor % (average)	Number and Output of Reactors
Canada	77.1	10 (5,960 MWe)	55.2	13 (7,865 MWe)
Japan	65.8	24 (16,766 MWe)	57.9	25 (17,432 MWe)
USA	55.3	73 (61,073 MWe)	53.4	77 (65,430 MWe)
West Germany	68.0	11 (10,248 MWe)	60.1	11 (10,248 MWe)
Sweden	68.3	9 (6,787 MWe)	53.2	10 (7,767 MWe)
France	53.0	28 (22,956 MWe)	52.9	30 (24,826 MWe)
Great Britain	51.4	23 (8,219 MWe)	52.5	23 (8,219 MWe)

Average Load Factors by Reactor Type (1982) (*Reactors with capacity greater than 150 MWe*).

	Average Load Factor	Number of Reactors	Total Capacity	Percent of Total	Average Size
PWR	60.0	103	85,343	59.0	829
BWR	61.2	56	41,559	28.7	742
PHWR	70.4	13	6,666	4.6	513
Magnox	53.0	26	8,527	5.9	328
AGR	58.5	4	2,640	1.8	660

Source: *Nuclear Engineering International.* August 1983, Supplement. Surrey, U.K.: Business Press International Ltd.

The mismatch between plans and performance in the past has also resulted in stockpiles of fuel materials. The fact that fuel-supply planning is conservative compared to actual operations—a practice that is justified by the large magnitude of reactor capital relative to fuel-cycle costs—means that other forms of flexibility are available as well. The reduced urgency of some consumers' needs may allow rescheduling, or even reassignment, of material by a fabricator, enricher, or other supplier in order to meet the needs of consumers whose fuel needs have not been met. Suppliers have also been conservative in their production planning. For example, in 1973 the United States required potential future enrichment customers to enter into contracts well in advance of reactor startup. Similar commitments were required of participants in fuel-cycle ventures in Europe.

Since actual deployment of reactors has not kept pace with the plans on which fuel commitments were made, surpluses of fuel have been accumulating. Many utilities in industrialized countries now hold three, four, or more years forward supply of nuclear fuel at various stages of processing; inventories are somewhat smaller in developing countries.

Near-term technical flexibility in the nuclear power industry protects countries from serious consequences in the case of brief, occasional interruptions in the supply of fuel. However, it will do little to increase actual assurance of supply if fuel procurement depends crucially on conditions within the markets where these goods are traded. The number of possible points of interruption, the concentration of supply in a few nations, the problems of restarting fuel-cycle flows after a disturbance, and the relatively high level of institutional intervention have undermined fuel security in the past, and may do so again.

MARKET STRUCTURE

Each of the supply stages in Figure 1–1 is part of a set of interlinked markets in nuclear materials and associated processing services.[3] Figure 1–2 shows the distribution (as of the end of 1982) among nations of uranium produc-

Figure 1–2.   Capacity Distribution for Nuclear Fuel Services.

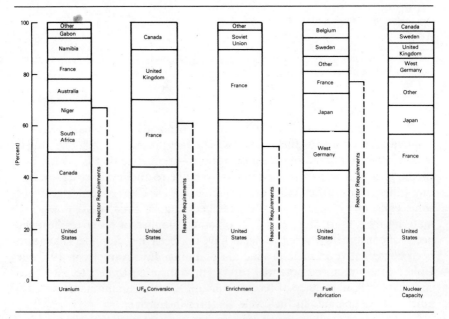

Source: Tables 1–3, 1–6, and author's estimates in Chapters 4 through 8.

Note: Capacity distribution is as of January 1983. Fabrication capacity is that for light-water reactors only. Reactor requirements shown are the percentages of capacity utilization required for reactor use (with appropriate lead times); additional capacity may have been used with the resulting material being added to inventories.

tion, conversion, enrichment, and fabrication capacities. Also shown is the distribution of reactors around the world and the demands they put on the fuel system, stage by stage. The various fuel-cycle steps are interdependent; for example, enrichment contracts often determine quantities of uranium procured and the timetable for fabrication. Uranium hexafluoride conversion capacities and fuel fabrication capacities are shown in Table 1–3. In what follows, we examine the structure of enrichment and uranium markets in general terms.

The process of securing fuel for the world's light-water nuclear power reactors involves contracting for uranium and for enrichment services, as well as for conversion and fabrication. Thus, utility fuel buyers must participate in several different international markets, each with its own imperatives and problems. Natural uranium trade is dominated by five key exporter countries: Australia, Canada, Niger, Namibia, and South Africa. Enrichment services are provided only by several highly industrialized countries or groups of countries: the United States, the Soviet Union, Urenco (a

Table 1–3. 1982 Uranium Hexafluoride Conversion Capacity (*MTU annually*)

	Existing	*Planned*
Allied Chemical (U.S.)	12,700	—
Kerr-McGee (U.S.)	9,090	—
British Nuclear Fuels Ltd.	9,500	2,000
Comhurex (France)	12,000	2,000
Eldorado Nuclear (Canada)	5,500	9,000
Other (Brazil, Japan, South Africa)	—	1,000
Total	48,790	14,100

Light-Water Reactor Fuel Fabrication Capacities (*MTU annually*)

United States	3,015
Japan	1,050
France	600
West Germany	1,050
Belgium	400
Sweden	400
Italy	200
Spain	200[a]
United Kingdom	125
Total	7,040

Note: Does not include fabrication for natural uranium reactors.

a. Under Construction.

Source: Nuclear Fuel Assurance Corp., March 1983 (Private Communication). End of year 1982 figures.

multilateral consortium of West Germany, the Netherlands, and Great Britain), and Eurodif (a consortium of France, Italy, Belgium, and Spain).

The policies of suppliers in both markets are complex and generally rigorous in commercial terms and, frequently, in political terms. Because of this and because of the technical factors described above, contracts for both uranium and enrichment services may extend for five, ten, or more years. Both markets are thus relatively inflexible in responding to changes in supply and demand conditions. But changes in demand expectations are a central fact of nuclear programs worldwide, and as a result, uranium contracting, enrichment contracting, and reactor fuel needs have often become mismatched. This mismatch has been aggravated by security-of-supply concerns on the part of consumer nations over access to enrichment supplies (in the early to mid-1970s) and then over access to uranium (in the mid- to late 1970s). In general, contracts for both uranium and enrichment worldwide exceed actual reactor needs, and growing pressure for both markets to readjust is leading to new market mechanisms.

The coupling of uranium and enrichment markets also creates special problems (and opportunities) for sellers of uranium or enrichment services. For example, commitments to enrichment contracts or equity shares in enrichment plants (where penalties for cancellations and readjustments may be high) can sustain demand for uranium even when reactor schedules slip. Similarly, long-term uranium commitments (either contracts or, especially, equity shares created by prepaid investments in mines and mills) can result in enrichment supply imperatives over and above expected reactor requirements. In addition, the market power exercised by some uranium suppliers may also allow them to enter the enrichment market as new suppliers. Such opportunities depend critically on conditions in both markets and on the detailed contractual positions of particular uranium producers and consumers in those markets. These conditions are treated in some depth in later chapters. We begin here with an overview of the structure of the uranium and enrichment markets.

Uranium

Uranium is a widely distributed element in the earth's crust but its discovery and commercial exploitation has been limited to a relatively small number of countries. A detailed assessment of the world uranium industry is a central theme of subsequent chapters. In this section, we present a simple overview of the basis for the international market. The production of uranium is concentrated in a relatively small number of countries. As shown in Figure 1-2, the United States, Australia, South Africa, Canada, Namibia, and Niger accounted for 90 percent of noncommunist output in 1982. Resources

Units and Definitions

Quantities of uranium and fuel-cycle services are frequently stated in units that are often not easily reducible to a common measure. For example, quantities of uranium may variously be described in terms of pounds of uranium oxide or kilograms of uranium hexafluoride. Throughout this book an effort is made to state quantities consistently in metric units and in terms of the uranium metal contained in various chemical compounds. We refer most often to metric tonnes of uranium (MTU), an approach that allows us to keep track of the critical ingredient in nuclear fuel no matter what chemical transformations have occurred. Conversion factors are as follows:

$$1 \text{ Kg U} \quad = \quad \begin{array}{l} 1.134 \text{ Kg UO}_2 \\ 1.179 \text{ Kg U}_3\text{O}_8 \\ 1.479 \text{ Kg UF}_6 \\ 2.600 \text{ lb U}_3\text{O}_8 \end{array}$$

$$1 \text{ MTU} \quad = \quad 1.300 \text{ STU}_3\text{O}_8$$

$$1 \text{ STU}_3\text{O}_8 \quad = \quad 0.769 \text{ MTU}$$

$$1 \text{ lb U}_3\text{O}_8 \quad = \quad 0.385 \text{ Kg U}$$

Quantities of enrichment services are usually stated in Separative Work Units (SWUs). An SWU is the measure of how much work must be done (in the thermodynamic sense) to enhance the concentration of the U-235 isotope in uranium. Quantities of enrichment are most correctly stated in mass units—the number of kilograms Separative Work, or metric tonne Separative Work (MTSWU)—though the mass label is frequently dropped and reference simply made to SWU. The enrichment process involves the input of several kilograms of natural uranium (as uranium hexafluoride containing 0.711 percent U-235) and the performance of a certain amount of Separative Work. What exits from the process is a stream of enriched material to be used to make fuel and a waste stream, or "tails," of depleted material. Typically, about 5.5 kilograms of natural uranium (contained in 8.1 kilograms of uranium hexafluoride) and 4.3 kg-SWU are used to produce one kilogram of uranium enriched to 3 percent U-235 and about 4.5 kilograms of tails material containing 0.20 percent U-235.

Quantities of uranium feed and enrichment services may be substituted for each other, within a range. Thus, for the same amount of fuel, one may reduce the amount of uranium sent to the enrichment plant if more enrichment is used, or one may increase the amount of uranium and reduce the amount of SWUs required. This decision is a function of relative prices for uranium and enrichment and of contractual and other factors, as described later in this chapter.

are similarly concentrated. Australia, Canada, Niger, South Africa, Namibia, and the United States together have nearly 85 percent of "reserves and resources" as estimated by the Organization for Economic Cooperation and Development (OECD).[4] While there is exploration in a large number of other nations and new output in perhaps ten countries, the total added to world production capacity is small (about 1,000 metric tonnes of uranium (MTU) annually compared to more than 38,000 MTUs of capacity in the traditional producer nations), and in most cases the material will be used domestically, rather than entering into international trade.

The United States and France are net importers of uranium, and will continue to be so in the future. Since they both import and export (with a net import balance), they represent opportunities for diversification of supply. However, if one is interested in internationally traded uranium, it is more informative to look at the level of concentration among net exporters. Here the level of concentration is even greater: if one sets aside the United States, France, and Gabon (which participates primarily in the French supply system), then Australia, South Africa, Canada, Namibia, and Niger account for 98 percent of remaining production.

A variety of firms and agencies participate in uranium production. First, governments are directly involved in resource exploitation. In Canada, for example, Eldorado Nuclear Ltd., a Canadian crown corporation, shares ownership of the Rabbit Lake and Key Lake deposits in Saskatchewan; the Saskatchewan Mining Development Corporation, a provincial government authority, has stakes in Key Lake, Cluff Lake, and Dawn Lake deposits and retains rights to future developments in the province. The South African government owns 13.5 percent of the Rössing operation in Namibia through its Industrial Development Corporation. France, Niger, and Gabon all have significant governmental interests in their domestic uranium industries. In exploration, state enterprises from both host and consumer governments play important roles. The trend in the 1980s seems to be one of increasing producer governmental involvement, especially in the richest deposits, often because of the resource rents involved, but also because of the presence of market and other risks.

A second group of participants consists of consumer utilities and consumer government agents. During the tight market of the 1970s, countries with major import requirements moved aggressively to explore abroad and to acquire interests in foreign uranium production ventures. Japan and France have been especially prominent, though Italy, Spain, and other countries have committed financial and technical resources. While some large utilities, or consortia, launched such efforts, the means of foreign involvement was often through state enterprises or private corporations acting with official backing. In the case of companies like these, it is often difficult to separate national interests from commercial motivations. Such activities were not limited to foreign utilities: during this same period, nearly half of the U.S. utilities with nuclear power plans reported some direct involvement in uranium production, though many of these utilities have now withdrawn from such ventures.

The third major group participating in uranium production consists of commercially motivated private companies. Outside the United States large companies dominate. The largest of these is the Rio Tinto Zinc group which has interests in several major producing nations: in 1982 the London parent company controlled nearly 12 percent of non-U.S. uranium production capacity and together with its affiliates nearly 18 percent. In Gabon and Niger, as in Metropolitan France, a group of private French companies work closely with French state enterprises. In the United States, where deposits are smaller on average, there has been an important fringe of smaller producers. Internationally, issues of scale and risk have led to numerous joint ventures.

In subsequent chapters, we trace the ownership of resources and production capacity in the major producer nations—Australia, Canada, Gabon, Namibia, Niger, and South Africa. It is possible to identify four categories of involvement: private companies based in the producer country, private companies based outside the nation (usually in a consumer nation), government agencies in the producer nation, and foreign government agents (usually consumer government agents). The results of this analysis are shown in Figures 1-3 and 1-4. Figure 1-3 shows the breakdown by category for production capacity as of the end of 1982 in each of the six major producer nations. As is evident, foreign companies or government agents play important roles in Canada, Gabon, Namibia, and Niger. Worldwide, private domestic firms control nearly 48 percent of production capacity; foreign private concerns control 27 percent; local governments more than 11 percent; and foreign governments nearly 14 percent.

Figure 1-4 shows the corresponding breakdown for control of reserves. Only well-defined deposits with identified ownership and some prospect of development are included. Note that prospectively exploitable reserves are greater for Australia than for Canada, the reverse of the situation for production. The reserve figure for Australia does not include the massive (1,000,000 MTU) deposit at Roxby Downs. Worldwide, the percentages for each type of ownership are not greatly different for reserves than for production capacity, though individual nations show significant shifts. In Australia, foreign private entities have a higher percentage of reserves than of existing production capacity, in part reflecting government ownership restrictions on ventures allowed to proceed to development. In Canada, there is an evident shift toward domestic government control. There is no major difference between production and reserves for the African producers. In a sense, Figure 1-4, which documents developable reserves, reveals the directions in which production capacity additions are likely to evolve.

Figure 1–3. Control of Production Capacity.

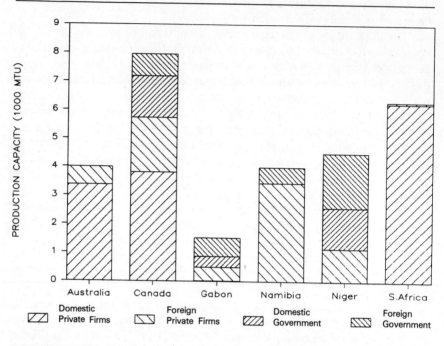

Note: Allocations shown are based on analyses of Chapters 5 and 6. In Australia, Nabarlek and Ranger are included; in Canada, Denison, Rio Algom, Rabbit Lake and Cluff Lake; in Niger, Somaire and Cominak operations; in Namibia, the Rössing mine; in Gabon, the four deposits actively exploited are included; and in South Africa, only those mines in operation as of early 1983 are included.

The issue of control is really somewhat more complex than these calculations suggest. Producer governments generally have greater influence over production decisions–and over price and other export conditions than equity shares alone indicate. Under weak market conditions this power can be used to restrain expansion and otherwise affect internal events. In sellers' markets, producer states are frequently in a position to impose price and nonprice conditions on exports and the external market. From the perspective of consumers, this power over the international market increases most when it is most feared—that is, when supply tightens and producers can demand both economic and political prices for fuel supply.

Uranium Prices and Contracting

For any commodity, the form of contracts and the ways in which prices are specified reflect the structure of the market and perceptions of risks. As

Figure 1-4.   Control of Uranium Reserves.

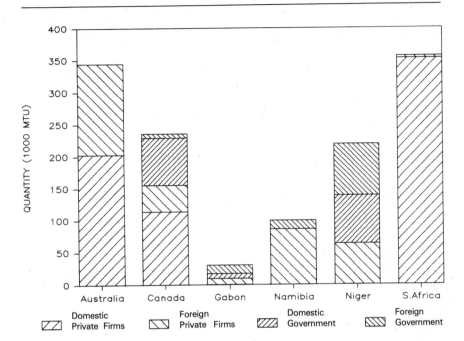

Note: Deposits included are those of Figure 1-3 plus Yeelirrie, Koongarra, Honeymoon, Beverley, Ben Lomond and Lake Way in Australia; Key Lake in Canada; and Arni, Abkorun, Imouraren and West Afasto in Niger. This figure does not include the large Roxby Downs deposit in South Australia, where British Petroleum—partially owned by the government of the United Kingdom—is a participant.

structure or perceptions change, so also do contracting mechanisms. Uranium has displayed a particularly exaggerated form of such change, being influenced not only by rapidly changing basic supply and demand outlooks but also by changes in enrichment and other linked markets and by a host of policy-related interventions into international trade. These changes are chronicled in subsequent chapters, leading up to a more detailed discussion in Chapter 9.

Information on uranium prices is difficult to develop. Not only are transactions usually secret, but there are really many different uranium prices, perhaps as many as there are contracts. Some pricing arrangements reflect consumer financial investments, others reflect the effects of exporter government price floors or price approval processes. Given these contractual complexities and lack of information, many analysts fall back on data series compiled by uranium brokers and based on what are essentially spot transactions or estimates of prices for such transactions. Nuexco, a firm

Table 1-4. Uranium Price Series 1950–1983.[a]

Year	Nominal Price	U.S. GNP[b] Deflator	U.S. Real Price	OECD[c] Exchange Rate	OECD Real Price
1950	9.11	52.65	17.30		
1951	10.10	57.01	17.72		
1952	11.28	57.57	19.59		
1953	12.35	58.80	21.00		
1954	12.27	59.58	20.59		
1955	12.25	60.65	20.20		
1956	11.51	62.43	18.44		
1957	10.49	64.65	16.23		
1958	9.45	65.79	14.36		
1959	9.12	67.55	13.50		
1960	8.75	68.56	12.76		
1961	8.50	69.22	12.28		
1962	8.15	70.48	11.56		
1963	7.82	71.46	10.94		
1964	8.00	72.57	11.02		
1965	8.00	74.13	10.79		
1966	8.00	76.58	10.45		
1967	8.00	78.53	10.19		
1968	6.50	83.99	7.74		
6/69	6.25	86.10	7.26		
12/69	6.20	88.62	7.00		
6/70	6.30	91.07	6.92		
12/70	6.15	93.03	6.61	N.A.	N.A.
6/71	6.05	95.70	6.32		
12/71	5.95	97.39	6.11		
6/72	5.95	99.42	5.98		
12/72	5.95	101.54	5.86		
6/73	6.50	104.75	6.21	100.00	6.21
12/73	7.00	108.74	6.44		
6/74	10.50	113.48	9.25	100.70	9.32
12/74	15.00	119.79	12.52		
6/75	22.00	124.44	17.68	98.34	17.39
12/75	35.00	128.99	27.13		
6/76	40.00	131.30	30.46	105.57	32.16
12/76	41.00	134.99	30.37		
6/77	42.25	139.01	30.39	103.31	31.40
12/77	43.20	143.24	30.16		
6/78	43.40	148.89	29.15	92.39	26.93
12/78	43.25	155.38	27.83	88.52	24.64
6/79	43.00	161.85	26.57	89.56	23.79
12/79	40.75	168.05	24.25	86.32	20.93
6/80	31.50	176.52	17.85	85.29	15.22
12/80	28.00	185.16	15.12	90.99	13.76
6/81	24.25	193.17	12.55	106.86	13.41
12/81	23.50	201.55	11.66	105.21	12.27
6/82	19.25	205.98	9.35	118.91	11.11
12/82	20.25	210.42	9.62	119.22	11.47
6/83	24.00	215.17	11.15	125.16	13.96
2/84	17.50	218.67	8.00	131.71	10.54

Table 1-4. *(Continued)*

a. 1950–1967 from USAEC purchases (ERDA. 1976. *Statistical Summary of the Uranium Industry*. Grand Junction, Colorado); 1968–1983 from Nuexco (NUEXCO Exchange Corporation. April 30, 1979. "NUEXCO Monthly Report to the Nuclear Industry." Menlo Park, California: Nuexco. No. 117. p. 1.4.) spot-market price (Exchange Value). Note that prices prior to 1970 reflected government purchase pricing for larger quantities rather than open-market conditions for spot purchases.

b. Corrected for inflation (deflated) by using the U.S. GNP Implicit Price Deflator (1972 = 100) for the United States (*Survey of Current Business*. October 1982, and various 1983 issues. Washington, D.C.: U.S. Department of Commerce, Bureau of Economic Analysis); and internationally by using the Multilateral Trade weighted exchange rate index (local currency/$) March 1973 = 100. Countries included in index are Canada, France, Italy, West Germany, Japan, the U.K., and others. (*U.S. Federal Reserve Bulletin*. Various issues, Washington, D.C.: U.S. Federal Reserve System, Board of Governors.) OECD exchange rates listed prior to 1978 (at June) are averages for entire year. Subsequent data are for actual month indicated.

c. Organization for Economic Cooperation and Development.

headquartered in the United States, and Nukem, with head offices in West Germany, both routinely publish such prices.

In Table 1-4, we have combined data series for early purchases by the U.S. Atomic Energy Commission (which set the purchase price) with more recent Nuexco price information (based on the "Exchange Value") to produce a time series of uranium prices, beginning in 1950. Table 1-4 also shows the effects of inflation and changes in world currency exchange rates. We have chosen to use the U.S. gross national product deflator, since this inflation index is commonly available and reflects general consumer price levels. It is also used as the index in a number of uranium contracts. Other choices for correcting for inflation—such as a non-residential structures index, an industrial commodities index, or an appropriate regional mining cost index—might also be used, depending on one's purpose. The series shown reflects real price experience in the United States. To give a general impression of the effect of varying currency exchange rates, we use a multilateral exchange rate index based on a weighted basket of OECD currencies.

As Table 1-4 reveals, spot uranium prices reached a peak (in nominal terms) in mid-1978. However, the real price (corrected for inflation) had evidently peaked, both in the United States and abroad, nearly two years earlier, in mid-1976. Prior to mid-1981, the real price of uranium to OECD nations was generally below that in the United States. However, increased strength in the U.S. dollar since that date has reversed this relationship, with many OECD nations now paying significantly more than the United States.

While Table 1-4 provides data that are adequate for some purposes, it does not fully reflect the complexity of uranium pricing. Prior to 1958, the U.S. Atomic Energy Commission (AEC) provided economic incentives to

U.S. producers and paid for roads and other infrastructural investments that do not show in the price series of Table 1–4. Outside the United States, the AEC appears to have paid prices above those indicated for the United States. Beginning in 1958, incentives in the United States were reduced, as were prices and contracted quantities. Outside the United States, the AEC announced a halt in further contracting in 1959 and stretched out future deliveries until 1967. However, average U.S. foreign procurement prices stayed above U.S. prices. This was not apparently the case with purchases by the United Kingdom and other buyers. For example, in 1962 the United Kingdom renegotiated a contract with Canada for deliveries between 1963 and 1971 for a price of $5.03 per pound. In 1964, when the U.S. domestic price was US$8.00 per pound, the United Kingdom solicited proposals in Canada to replace output from a failed mine at C$4.18 (US$3.87) per pound.[5]

Since the late 1960s, large quantities of uranium have been sold in commercial markets at prices above and below those shown. One would expect that the average price paid by utilities would not vary over time as much as the spot price. While comprehensive world data are not available, this is borne out by price data reported by utilities and other users in the United States. Annually since 1975, the Department of Energy has collected information on average uranium prices paid for deliveries in the prior year.[6] These prices are shown in Table 1–5, along with inflation-corrected prices. For convenience, comparison is also made with the spot prices from Table 1–4.

Table 1–5.   Annual Average Prices Paid by U.S. Utilities (*Dollars per pound*).

Year	Average Nominal Price	Average Real Price	Nominal Spot Price	Real Spot Price
1973	7.10	6.78	6.79	6.48
1974	7.90	6.96	10.83	9.54
1975	10.50	8.44	24.00	19.92
1976	16.10	12.26	38.60	29.40
1977	19.75	14.21	42.15	30.32
1978	21.60	14.51	43.28	29.08
1979	23.85	14.76	42.33	26.15
1980	28.15	15.95	33.42	18.37
1981	32.20	16.67	25.25	13.07
1982	38.37	18.63	21.00	10.19

Sources: Data for 1973 and 1974 from *Statistical Data of the Uranium Industry*, GJO-100 (1979); other data from *Survey of United States Uranium Marketing*, April 1976, May 1977, May 1978, August 1979, July 1980, June 1981, and September 1983, U.S. Department of Energy. Spot prices, from Table 1–4, are averages for the year. All prices are deflated according to mid-year index.

Enrichment

Enrichment supply arrangements today, and into the future, are largely the result of decisions and commitments made up to a decade ago. Concerns in the early to mid-1970s about future access to enrichment supplies led many consumer nations to commit either to contracts with the United States (which at the time held a near-monopoly on enrichment) or to new joint ventures in Europe to build enrichment capacity. During the time it took to construct these new facilities, a number of European nations also contracted with Techsnabexport, the Soviet Union's enrichment export agency.

Sales to European nations by the Soviet Union were the first step in the erosion of the U.S. monopoly position; in the late 1970s, the Soviet Union provided enrichment services to Western Europe comparable to those from the United States. Subsequently, two European enrichment consortia entered the market. Urenco, a tri-national consortium of British, Dutch, and West German interests, made its first commercial deliveries in 1976 and has plans to expand its centrifuge enrichment capacity through the 1980s, Eurodif, a consortium involving France, Belgium, Spain, Italy, and originally Iran, made its first commercial deliveries in 1979 and quickly increased its gaseous diffusion capacity to nearly half that operated by the United States at that time.

In addition to these ventures, a number of others were announced in the mid-1970s when it appeared that enrichment capacity might be inadequate in the next decade. The Erodif partners planned a new venture, Coredif, a 10.8 million SWU per year plant with ownership shares somewhat different than those of Eurodif. Subsequently, surplus enrichment capacity, even at the Eurodif facility, has effectively eliminated incentives to build Coredif. South Africa's Uranium Enrichment Corporation (UCOR) announced plans to build a commercial facility using a stationary-wall centrifuge process, an effort to gain autonomy in enrichment. Brazil's Nuclebras, with the assistance of West Germany, planned a 0.2 million SWU demonstration facility in the mid-1980s using German Becker-nozzle technology, though there appear to be technical and other difficulties with this plan. Japan's Power Reactor and Nuclear Fuel Development Corporation (PNC) is considering expansion from a current pilot centrifuge plant to a facility with one million SWUs or more annual capacity by the mid- to late 1980s. At various times, interest in acquiring enrichment capability has been expressed by Australia, India, Iran, Portugal, Pakistan, Sweden, and Zaire.

As a result of new development initiatives undertaken when it appeared that enrichment demand would rapidly outgrow existing capacity, and as a result of the desire of various nations and groups of nations to attain greater autonomy, enrichment capacity expansion has been greater than is likely to be needed. Table 1–6 shows capacity development for major ventures and announced plans for future expansion.

Table 1–6. Announced Plans: Enrichment Capacity Expansion Worldwide[a] (*1,000 MTSWU*).

Year	U.S. DOE[b] GDP	GCEP	AVLIS	EURODIF[c] GDP	URENCO[d] GCP	USSR[e] GDP	Total
1978	17.2	—	—	—	—	3.2	20.4
1979	17.2	—	—	2.2	—	3.2	22.6
1980	17.2	—	—	6.0	0.4	3.2	26.8
1981	23.5	—	—	8.5	0.4	3.2	35.6
1982	25.0	—	—	10.8	0.7	3.2	39.7
1983	27.3	—	—	10.8	1.0	3.2	42.3
1984	27.3	—	—	10.8	1.3	3.2	42.6
1985	27.3	—	—	20.8	1.5	3.2	52.8
1986	27.3	—	—	10.8	2.0	3.2	43.3
1987	27.3	—	—	10.8	2.5	3.2	43.8
1988	27.3	0.4	1.0	10.8	3.0	3.2	42.7
1989	27.3	1.1	1.0	10.8	3.5	3.2	46.9
1990	27.3	3.0	1.0	10.8	4.0	3.2	49.3
1991	27.3	5.1	1.0	10.8	4.0	3.2	51.4
1992	27.3	7.1	1.0	10.8	4.0	3.2	53.4
1993	27.3	9.4	1.0	10.8	4.0	3.2	55.7
1994	27.3	11.5	1.0	10.8	4.0	3.2	57.8
1995	27.3	12.7	9.8	10.8	4.0	3.2	67.8
1996	27.3	13.2	9.8	10.8	4.0	3.2	68.3
1997	27.3	13.2	9.8	10.8	4.0	3.2	68.3
1998	27.3	13.2	9.8	10.8	4.0	3.2	68.3
1999	27.3	13.2	9.8	10.8	4.0	3.2	68.3
2000	27.3	13.2	9.8	10.8	4.0	3.2	68.3

a. Capacities prior to 1982 based on reported construction; estimates for 1982 and subsequent years are based on announced plans, as reported by Nuclear Assurance Corporation ("Worldwide Enrichment Report," Internal Document, March 1983). GDP = Gaseous Diffusion Plant; GCEP = Gas Centrifuge Enrichment Plant; AVLIS = Atomic Vapor Laser Isotope Separation Plant.

b. Department of Energy diffusion plants at Oak Ridge, Tennessee; Paducah, Kentucky; and Portsmouth, Ohio, and laser isotope-separation facilities currently planned at Oak Ridge.

c. Eurodif gaseous diffusion facility at Tricastin, France.

d. Urenco centrifuge facilities at Almelo, The Netherlands; Carpenhurst, United Kingdom; and Gronau, West Germany.

e. Actual Soviet capacity is believed to be about 10,000 MTSWU, not all of which is available for export. The quantity shown is the 1980 export level: additional quantities might be available if market conditions warrant.

For three of the principal suppliers, the United States, the Soviet Union, and Eurodif, enrichment services are currently provided by gaseous diffusion plants which have large economies of scale and (generally) long-term electric power commitments. The Urenco facility uses centrifuges that can be added in smaller increments as demand increases. The capacity shown for the U.S. diffusion plants reflects the cascade upgrading and cascade improvement programs. The U.S. Department of Energy is committed to deployment of centrifuge capacity of 2.2 million SWUs and plans also include additional increments. Laser isotope-separation technologies are still in the research and development process; the capacity shown assumes success in this effort and commercial deployment at high levels in the late 1990s. Budgetary and other pressures seem likely to force alterations in current U.S. programs. The announced expansion of the Urenco centrifuge capacity reflects the construction of new facilities in addition to those currently under construction at Almelo. The capacity shown for the Soviet Union is that which historically has been available for export through the agency Techsnabexport. Though exact figures are not publicly known, total Soviet capacity is reported in many places to be about 10,000 MTSWU annually.

Of the announced or existing capacity shown in Table 1-6, not all will be used for commercial purposes, or even constructed. Current world capacity is about three times as much as required to meet short-run commercial requirements (see Chapter 4), and virtually all facilities are operating below design capacity. In 1982, the U.S. plants operated at an annual capacity of only about 9.8 million SWUs[7]; the Soviet agency, Techsnabexport, exported about 2.1 million SWUs to Western nations; and Eurodif produced about 6.4 million SWUs.[8] Even these levels exceeded reactor requirements due to overcontracting and financial and other commitments to facilities.

This oversupply situation seems likely to continue for some time. Figure 1-5 shows historic enrichment deliveries to utilities and agents outside of the United States and recent estimates of likely future deliveries, based on contractual and other commitments.[9] For the United States, export quantities shown for 1983 and subsequent years are scheduled deliveries expected as of early 1983, taking into account expected deferrals; additional deferrals and cancellations are possible. Eurodif deliveries are somewhat more rigidly specified by equity investments and other prior commitments, though Italy and Spain are currently taking electric power deliveries in place of enrichment services (electrical generation capacity had been constructed in conjunction with the Eurodif facility, and electricity makes up a significant component of enrichment cost for diffusion facilities). The deliveries shown for Techsnabexport are those for known contractual deliveries to Western nations. There is considerable economic pressure for deferral of Soviet deliveries that could result either in price cuts or in delays in deliveries or both. Soviet decisions are likely to depend on prospects for future sales. As indicated in Figure 1-5, Techsnabexport deliveries fall off rapidly after 1990 and will result in a significant drop in Soviet foreign-exchange earnings if new contracts are not written.

Figure 1–5. World Enrichment Deliveries.

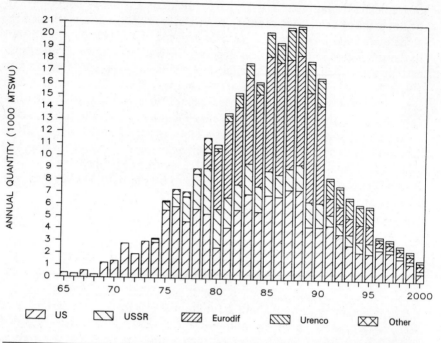

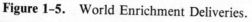

Sources: Lawrence Livermore Laboratory. April 1978, October 1980, and January 1983. "Status of Foreign Uranium Enrichment Activities." U.S. Department of Energy; Author's estimates. Historic U.S. exports and future estimates from U.S. Department of Energy, Oak Ridge Operations Office. June 1983. Reports 1475 and 1482. Figure does not include 10,020 MTSWU advance sale by the United States to Japan in 1973.

Note: Deliveries shown are to nations other than the United States and the centrally planned economic areas. Industry convention is to refer to the non-centrally-planned economies as World Outside Communist Areas (WOCA).

Comparison of Table 1–6 and Figure 1–5 shows that existing and announced capacity will be more than twice the currently expected delivery levels even near the end of the decade. Given this continuing oversupply situation, it seems unlikely that new capacity expansion will proceed as rapidly as announced plans suggest, unless deployment of new technologies offers the potential for significantly reducing enrichment costs and thus displaces existing diffusion plant capacity.

Financial and other commitments to enrichment—especially those associated with the gaseous diffusion plants—have made it difficult for consumer utilities to adjust enrichment delivery commitments, and thus they remain obligated to deliver uranium feed. As reactor capacity expansion has been delayed, relative to what was expected when commitments were made to enrichment services, there has been a tendency for enrichment plant delivery schedules to replace reactor feed requirements as the dominant factor in setting uranium demand, at least on a four- or five-year horizon. As discussed below, supply and demand mismatches have resulted in an active secondary market in both uranium and enrichment services, first in the United States and now spreading abroad.

Out of all enrichment commitments by non-U.S. utilities (from the U.S. and non-U.S. suppliers), some can be adjusted within the span of a single year (10 to 15% of total non-U.S. utility commitments over the decade), some are flexible beyond a period of four or five years (less than 20%), and most are relatively inflexible (about 70%). Substantial additional changes are, of course, possible but at the cost of termination penalties for the United States and the Soviet Union or reduced utilization of Eurodif and Urenco facilities (resulting in reduced, or perhaps negative, return on investment for the equity participants). For some groups of utilities it is possible to pool enrichment contracts—the flexible with the inflexible—and thus obtain greater flexibility than might be possible for a single utility. There is also an active secondary market—in which utilities and agents resell enrichment services or enriched material—that began first in the United States and has now spread to Eurodif and other suppliers.

By the late 1980s, however, enrichment-contract arrangements are likely to become more flexible due to the market pressure of overcapacity in what is increasingly a competitive market. This pressure will also affect prices for enrichment. Investments in diffusion plants are sunk-costs, and these investments were often made by governments or encouraged and backed by governments. Whether and how these sunk-costs are recovered is a question that can be resolved in many different ways, depending on domestic and international political factors. Thus while operating costs (especially electrical power supply) for diffusion plants are admittedly high, there may be considerable flexibility in pricing. New enrichment technologies may also reduce costs, though the circumstances under which major new investments might be made remain unclear.

Table 1-7 shows the enrichment pricing history for the United States for two common forms of enrichment contracts: Requirements contracts and Long-Term Fixed Commitment (LTFC) contracts. Requirements contracts are more flexible and thus carry higher prices than LTFC contracts. Also shown is the price adjusted for the effects of inflation. This adjustment is shown for the Requirements contracts—the longest time-series available for world enrichment contracts. As is evident, the *real* price (corrected for inflation) for enrichment has not increased greatly over the past thirty years, though it did drop significantly in the early 1970s.

The U.S. prices are comparable to those charged by other suppliers,

Table 1-7. United States Enrichment Prices (*US$/Kg.-SWU*).

Year	Requirements	LTFC	Real Price[a]
11/56	37.29	—	58.55
07/61	36.50	—	52.27
07/62	30.00	—	42.57
01/68	26.00	—	32.04
09/71	32.00	—	33.15
08/73	38.50	36.00	36.14
12/74	47.27	42.10	39.48
12/75	59.80	53.35	46.36
08/76	65.83	59.05	49.54
07/77	67.58	61.30	48.62
07/78	83.15	74.85	55.85
07/79	95.09	88.65	58.75
07/80	106.61	98.95	60.40
01/81	110.75	110.00	58.29
11/81	127.59	130.75	63.30
07/82	141.15	130.75	68.53
12/82	149.85	138.65	71.21
11/83	149.85	138.65	69.26

[a]Requirements contract price adjusted by U.S. GNP deflator (1972 = 100).

Source: *Nukem Market Report.* Various issues. Hanau, Federal Republic of Germany: NUKEM GmbH.

with small differences depending on market conditions (in early years, the strong market position of the United States required other suppliers—such as the Soviet Union—to offer slightly better prices) and on exchange rates. For example, in 1982 the strength of the U.S. dollar pushed the effective price of U.S. enrichment for European and Japanese customers higher than the equivalent prices of other suppliers. As noted above, the enrichment-market should be quite competitive over the next decade, and it is likely that real prices will actually decline under such competitive pressures.

FUEL-CYCLE ECONOMICS

Uranium and nuclear fuel services represent only a small part of the cost of nuclear power generation, though this fraction depends on how one does the calculation and what one assumes. In general, capital expenditures for nuclear power plants may occur over a period of ten or more years and be recovered over perhaps thirty or more years. There are thus not only long time horizons that make the accounting complicated but also significant uncertainties. The latter include future inflation and interest, cost escalation in various inputs to construction, safety modifications during construction, changes in rate regulation, and other factors. Not surprisingly, it is possible (indeed likely, in such a contentious area) to get very different estimates for capital costs and for their effect on electricity prices.

Similarly, uranium and fuel-cycle services require considerable lead times, and because the fuel is used over a period of time, one must be concerned with the effects of financing, cost escalation, inflation, and other factors. As a result, nuclear fuel is usually capitalized, rather than being expensed like other fuels (such as oil) that are bought and burned in relatively close temporal proximity.

Many studies estimate the contribution for nuclear fuel to nuclear generation costs, usually by computing a "levelized" fuel cost: the average cost (in constant dollars of some specified year) of the fuel used over the life of the reactor. This calculation is useful for comparing costs that are otherwise hard to compare due to variations over time. The results range from perhaps 10 percent of total cost to more than 30 percent, depending on the assumptions made about future escalation rates for capital, fuel, and operating and maintenance expenses. However, the levelized-cost calculation requires assumptions about prices and other factors for quite a distance into the future, and these assumptions can obscure some simpler issues that are deserving of attention.

To look at the components of fuel cost, we remove the time variation by simply assuming that uranium is bought, converted, enriched, and fabricated all at one point in time. Table 1–8 lists recent prices (mid-1982) for various components of fuel cost. Two cases are considered. In the first, it is assumed that enrichment is conducted such that the tails assay is 0.20 percent. In this case, enrichment is a larger factor in the cost of fuel than is uranium (55% compared to 26%). In the second case, the enrichment tails assay is set at 0.33 percent. Here, uranium services increased (to about 35% of fuel cost) and enrichment services decreased (to about 44% of fuel cost), compared to the case of 0.20 percent tails assay. The result is that the total cost of fuel is reduced.

Figure 1–6 illustrates the more general point. For any relative prices of uranium (including conversion cost) and enrichment, there is an economically *optimal* tails assay that minimizes the total cost of nuclear fuel. This fact has several implications for uranium and enrichment demand and market conditions.

Expectations about uranium prices have changed radically over the past decade. In the early to mid-1970s, it was generally expected that uranium prices would rise rapidly, more rapidly than enrichment prices. As prices rose, it would be attractive to operate enrichment plants at lower tails assays—in effect, substituting enrichment for scarce uranium. As will be discussed in Chapter 2, there were also other factors that motivated setting contractual tails assays at relatively low levels. As a result, utilities con-

Table 1-8.   Components of Fuel Cost[a] (*$US/Kg. U*).

At 0.20 Percent Tails Assay

	Price	*Units Req.*	*Cost*	*Percent*
Uranium	52.00	× 5.479 =	284.90	26
Conversion	6.60	× 5.479 =	36.16	3
Enrichment (per SWU)	138.65	× 4.306 =	597.03	55
Fabrication	170.00	× 1 =	170.00	16
Total			1,088.09	100
Contribution[b] To Generation Cost (mills/KW-hr)			4.53	

At 0.33 Percent Tails Assay

	Price	*Units Req.*	*Cost*	*Percent*
Uranium	52.00	× 7.008 =	364.42	35
Conversion	6.60	× 7.008 =	46.25	4
Enrichment (per SWU)	138.65	× 3.229 =	447.70	44
Fabrication	170.00	× 1 =	170.00	17
Total			1,028.37	100
Contribution[b] To Generation Cost (mills/KW-hr)			4.29	

a. Fuel material enriched to 3 percent U-235 by weight. Conversion losses (perhaps 0.5%) are ignored. Carrying costs might add 50 to 100 percent to costs shown, depending on lead times.

b. Assumes burnup (total output over the life of the fuel) of 30,000 MWd(thermal)/MTU and reactor thermal efficiency of 33 percent. A Mega-Watt-day is a measure of power plant output, expressed in thermal (heat) or electrical units. One MWd(thermal) of output, at 33 percent efficiency of conversion to electricity, yields about 8,000 kW-hr. of electricity.

tracted or invested in enrichment plants (through their governments or other agents) for large quantities of enrichment services, with the implicit assumption that the desired tails would be in the vicinity of 0.20 percent.

At times, uranium and enrichment prices were in the right proportions (at a ratio of about 1.3) to make purchases at 0.20 percent assay economically optimal. But, more often, relative (spot) prices—and price expectations—have been such as to favor other tails assays. Taking into account the evolution of prices for both uranium and enrichment, it is possible to compute the change in optimal tails assay over time. In Figure 1-7, we show the results of this calculation using the sequence of spot uranium prices shown in Table 1-4 and LTFC enrichment prices shown in Table 1-7.

In using the spot price for uranium, we are essentially examining the optimal tails assay for the marginal unit of nuclear fuel. As is evident, the optimal tails assay at the margin has risen significantly since 1978. However,

Figure 1-6.   Optimal Tails Assay as Function of the Ratio of Uranium Feed and Enrichment Prices.

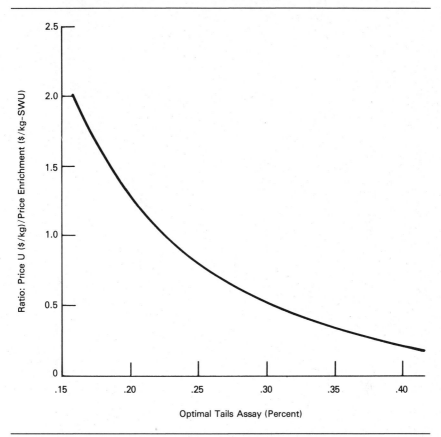

Note: Feed price is total for uranium plus price of conversion to uranium hexafluoride, both expressed in terms of uranium metal contained in uranium oxide or hexafluoride. Details of the calculation of the optimal tails assay are given in Appendix B.

one obtains quite a different result if one uses *average* delivered uranium prices, such as those in Table 1-5. In this case, the optimal tails assay for the majority of fuel (for U.S. utilities at least) has been remarkably constant— just above 0.25 percent—for much of the last decade. Changes in enrichment prices have just compensated for changes in uranium prices in such a way as to keep the optimal tails assay constant for the average fuel reload. Indeed, the optimum shown is very close to the assay at which U.S. enrichment plants have operated over many years, though it is above the tails assay at which contracts have usually been written.

Figure 1-7.   Changes In Optimal Tails Assay.

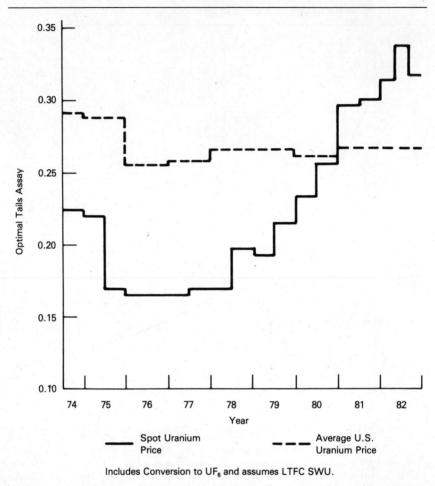

Includes Conversion to UF$_6$ and assumes LTFC SWU.

Note: Separate calculations are made of optimal tails assay in enrichment for two different uranium price series. The first—the more variable of those shown—is based on spot prices from Table 2-4. The second is based on the average price reported each year by U.S. utilities. Both calculations assume reported fabrication prices and enrichment prices (under LTFC contracts) in the year indicated.

Under market surplus conditions, where buyers have contracted for enrichment services greater than what would be needed to minimize fuel costs, there may be incentives to resell uranium or enrichment services, or both, or to pay cancellation penalties. In evaluating such incentives it is appropriate to use spot prices. We first estimate what the resale value of enrichment services might be or, alternatively, how much buyers should be willing to pay to reduce their enrichment commitments. Again, we simplify the calculation to avoid unnecessary assumptions. Suppose a utility contracted for enrichment services under the assumption that the optimal (marginal) tails assay would be 0.20 but that the optimal tails assay turns out to be 0.33 percent (for example, if the utility is able to buy uranium at low spot prices). Using the price parameters of Table 1–8, and assuming that the utility can adjust the tails assay in enrichment, we can see that the utility should be willing to buy 1.53 kg more uranium and take 1.08 less SWU to achieve a cost savings of about $60 on each kilogram of fuel.

However, the utility may be committed to take the extra 1.08 SWU at a price of $138.65 per SWU. In this case, the utility should be willing to sell the extra SWU at any secondary market price greater than $83 per SWU. Alternatively, the utility should be willing to pay a termination penalty of up to about $56 per SWU in order to achieve some net cost reduction. Of course, this example is idealized. Enrichment contracting and selection of trails assays are not as flexible as assumed; there are significant time lags in purchase commitments (which imply carrying costs); and average uranium prices for the utility are likely to be higher than the spot price due to long-term contractual obligations for uranium.

At present, cancellation penalties are smaller than indicated and discounts for resale SWUs are generally smaller than shown with (discounted present value) secondary market prices closer to $100 per SWU.[10] In the longer term, there are clearly strong economic pressures, comparable to those calculated, that will affect enrichment prices in both primary and secondary markets, as long as uranium prices stay low relative to enrichment prices. However, if adjustments in enrichment commitments are made, uranium demand will go up as tails assays increase, with a consequent upward pressure on uranium prices. Conversely, common efforts on the part of enrichers to require lower tails assays would further depress uranium prices.

INSTITUTIONAL FRAMEWORK

Nuclear trade takes place within a system of unilateral, bilateral, and multilateral conditions, agreements, and constraints. Although international commodity trade is often subject to such involvements, the rules governing nuclear fuel are particularly complex. Each fuel-cycle step can occur in a different country and under different legal and political conditions. Moreover, governments have a long history of involvement in nuclear activities and are often responsible for research and development, finance, regulation, and export promotion. The result is a set of political restraints and interventions that have a considerable effect on the supply of nuclear fuel.

Since supplies of uranium and enrichment have been concentrated in a few countries, these suppliers have sometimes been in a position to impose political as well as market conditions on the export of fuel. In Chapter 2, we review the development of the international institutional and policy framework within which nuclear fuel trade takes place. To prepare for that discussion, it is useful to look briefly at the international structure within which nations with nuclear power programs or industries operate, with special attention to the International Atomic Energy Agency (IAEA), the Non-Proliferation Treaty (NPT), the European Atomic Energy Commission (Euratom), and the London Suppliers' Group.

The Nuclear Non-Proliferation Treaty

The NPT, which went into effect in 1970, contains two basic obligations, one attaching to nuclear-weapon states and the other to non–nuclear-weapon states. Each nuclear-weapon state undertakes not to transfer nuclear weapons or control over those weapons, directly or indirectly, and agrees not to assist, encourage, or induce any non–nuclear-weapon state to manufacture or otherwise acquire nuclear weapons or control over them (Article I). Each non–nuclear-weapon state undertakes not to receive nuclear weapons or control over them, not to manufacture or otherwise acquire nuclear weapons, and not to seek or receive any assistance in their manufacture (Article II).

Under the safeguards provisions of Article III, each non-nuclear-weapons party to the treaty is obligated to apply IAEA safeguards to *all* nuclear facilities. Each party to the treaty also undertakes not to export fissionable material—or equipment for the use, processing, or production of fissionable material—unless IAEA safeguards are applied.

In complement to these commitments, all parties to the NPT "undertake to facilitate, and have the right to participate in, the fullest possible exchange of equipment, materials and scientific and technological information for the peaceful uses of nuclear energy" (Article IV—see Appendix A). This obligation has generally been interpreted as flowing from the nuclear-weapon states to the non–nuclear-weapon states.

However, there has been difficulty in agreeing on which materials and technologies are appropriate to the "peaceful uses" criterion. It has generally been the policy of the United States—and more recently of other suppliers—that proliferation-sensitive technologies, such as reprocessing or enrichment, or material such as plutonium, are not automatically included under the NPT obligation. Because of its fundamental importance to international nuclear trade, the text of the Non-Proliferation Treaty is included in Appendix A.

With the exception of France and Spain, all major industrial countries have signed or ratified the NPT. A number of other countries have not done so—among them are Argentina, Bangladesh, Brazil, Chile, India, Israel, South Korea, Pakistan, and South Africa. As noted earlier, however, even in these countries many civilian nuclear power facilities are under IAEA safeguards. In return for assistance in meeting nuclear energy needs, the customer state accepts the intrusion of safeguards on its sovereignty.

IAEA Safeguards

The International Atomic Energy Agency, established in 1957, serves a number of functions, including research, education, and nuclear promotion. But for the purposes of this discussion, the most important aspect of the agency is the nuclear safeguards system that it administers. The IAEA system interfaces with a number of national control systems and with the multination system of the Euratom nations. The IAEA implements the full-scope safeguards accepted by signatories to the NPT, though it also provides safeguards for selected facilities and materials transferred to non-NPT signatory nations.

The objective of the safeguards system is to provide "timely detection of diversion of significant quantities of nuclear material from peaceful nuclear activities to the manufacture of nuclear weapons . . . and deterrence of such diversion by the risk of early detection."[11] The safeguards are based on a system of materials accountancy that attempts to strike materials balances for various facilities, accountancy regions, and for international flows. The accountancy system is backed up by on-site inspection of a nation's accounting records and of the safeguarded facilities themselves. The IAEA has also taken an important role in advising governments on procedures for physical security of nuclear materials.

Under the terms of the Non-Proliferation Treaty (Article 3, Section 2), parties to the treaty agree not to provide fissionable material or equipment "designed or prepared for the processing, use or production of special fissionable material" unless safeguards are attached. Since the treaty does not specify in detail what these materials and equipment are, and since technological change may require adaptation over time, it was found necessary to implement a process that would define, and extend when necessary, lists of items requiring safeguards. Since the NPT went into force, such a "trigger list" has been provided by a multination committee, now known as the Zangger Committee.

This list is published as Information Circular 209 of the IAEA and includes materials and equipment whose transfer to non–nuclear-weapon states should involve formal government assurances concerning non-explosives use, physical protection, IAEA safeguards (including safeguards on facilities constructed using transferred technology), and the security of an IAEA safeguards presence in the recipient country. This trigger list initially included equipment and know-how that might be directly useable in weapons programs. Additions dealing with heavy water facilities were made in 1978 and discussions of adding items relating to centrifuge enrichment technology were conducted in 1982 and 1983. The trigger list is useful not only in making sure that transfers carry with them safeguards, but also in providing guidance to nuclear supplier nations and to their nuclear-exporting industries.

The detailed arrangements for participation in the IAEA system by a nation, or by a group such as Euratom, are negotiated case by case with the IAEA.[11] It is important to note that national safeguards systems vary considerably and that some of a nation's facilities may be under safeguards while others are not. For example, almost all major commercial nuclear power plants fall under the system (some at the insistence of the supplier country), but such involvement by non-NPT signatories in the safeguards system does not necessarily imply a commitment to subject all nuclear facilities to international surveillance.

Euratom

Established by the Treaty of Rome in 1957, Euratom was designed to serve the collective interests of the European nations[12] in competition with the United States. Originally, the treaty called for a supply agency with exclusive rights to contract for nuclear materials within and outside the European Community. To do so:

> An Agency shall be constituted, having a right of option on all ores, source materials and special fissionable materials produced in the territories of Member States and having the exclusive right of concluding contracts relating to supplies of ores, source materials and special fissionable materials coming from inside or from outside the Community.

> The Agency shall not make any discrimination between users based on the use they intend to make of the supplies requested unless such use is unlawful or is found to be contrary to conditions laid down by suppliers outside the Community in respect of the particular delivery concerned.[13]

Drafted in the atmosphere of the Suez crisis, the exclusive trade provision was meant to prevent discrimination in access to fuel supplies (enrichment or uranium), which might occur with separate bilateral arrangements, and to ensure equal advantage to Euratom nations in a supply crisis.

The exclusive trade function was brought into question in the mid-

1960s when France sought nuclear fuel supplies outside the Euratom supply channels. In 1971, France unilaterally arranged the purchase of enrichment services from the Soviet Union, an action ruled against by the European Court of Justice but with little effect. The function of the Supply Agency has remained an issue within the Community. In some situations, Euratom has provided a convenient mechanism for negotiation (or resistance to negotiation) of non-proliferation and other trade conditions, but its role as a fuel-supply channel has increasingly come into question. Active discussions about the role of the Supply Agency took place during 1982 and may eventually be followed by explicit changes in the Euratom Treaty.

From the beginning, the Euratom agreement provided for a free flow of material and information among members. Euratom also has its own internal safeguards system which is coordinated with the accountancy framework of the IAEA. In recent years, difficulties have arisen because some supplier states have insisted on acceptance of IAEA safeguards or other conditions on individual Euratom nations, resulting in conflicting safeguards requirements and European resistance to this intrusion on regional sovereignty arrangements. In particular, the Euratom principle of free flows of material among members clashes with "prior approval" clauses for retransfers in some supplier contracts (discussed below). As of the end of 1982, new Euratom agreements with Canada and Australia had been reached, either directly or indirectly, through bilateral agreements with Euratom nations. However, Euratom has thus far resisted U.S. demands for renegotiation of its bilateral agreement.

Suppliers' Group

With the growth of nuclear technology commerce in the 1970s came an increased concern about the transfer of "sensitive" technologies—technologies such as reprocessing and enrichment that would facilitate nuclear-weapons development in non-weapons nations. An important forum for the discussion of export policy issues was the London Suppliers' Group, an initially secret group of participants from nuclear exporting countries that began meeting in London in 1975. The initial members were the United States, the Soviet Union, the United Kingdom, France, West Germany, Canada, and Japan; other countries, including Belgium, Sweden, Italy, the Netherlands, East Germany, Czechoslovakia, and Poland, were participating by early 1977.

Participants have taken the position that the results of their discussions are not binding agreements but rather, to the extent that common positions emerge, a series of unilateral policy actions taken by individual governments. Such a position was made necessary by the legal restrictions of the

Euratom Treaty (which requires that all nuclear agreements entered into by members be approved by the Euratom Commission) and by a desire not to be regarded as a suppliers' cartel. The suppliers' group has not entirely escaped the latter interpretation, however, particularly in the developing world.

The suppliers' group has agreed to increasingly restrictive common export conditions on nuclear technology trade and has provided an opportunity to discuss nonproliferation issues. Consultation with the group is required for exceptions to the agreement. Suppliers agree to exercise restraint in the transfer of sensitive facilities, technology, and weapons-usable materials; multinational arrangements are to be preferred. Suppliers are also urged to make provision for subsequent "mutual agreements between the supplier and the recipient of arrangements for re-processing, storage, alteration, uses, transfer or retransfer of any weapons-usable material involved."[14] The guidelines also call for continued consultation between suppliers on the implementation of the guidelines, on the existence of possible violations of agreements, and on responses to such violations. Because these guidelines are not commonly available, they are reproduced in Appendix A.

REFERENCES

1. Neff, Thomas L., and Henry D. Jacoby. 1979. "Supply Assurance in the Nuclear Fuel Cycle." *Annual Review of Energy* 4:259–311.
2. OECD Nuclear Energy Agency. 1978. "Nuclear Fuel Cycle Requirements." Paris, France: Organization for Economic Cooperation and Development.
3. Except where otherwise noted, summary information used in this chapter is developed more fully in subsequent chapters, where detailed references are given.
4. OECD Nuclear Energy Agency, and the International Atomic Energy Agency. February 1982. *Uranium: Resources, Production and Demand.* Paris, France: Organization for Economic Cooperation and Development.
5. U.S.-Canadian exchange rates from *Standard and Poor's Statistical Service.* 1983. New York: Standard and Poor's Corporation.
6. U.S. Energy Research and Development Administration, Division of Nuclear Fuel Cycle and Production. April 1976. *Survey of United States Uranium Marketing Activity.* ERDA 76-46.

   U.S. Energy Research and Development Administration, Division of Uranium Resources and Enrichment. May 1977. *Survey of United States Uranium Marketing Activity.* ERDA 77-46.

   U.S. Department of Energy. May 1978, August 1979, July 1980, June 1981, and July 1983. *Survey of United States Uranium Marketing.*

7. Totals are reported monthly in: *Nuclear Fuel.* Various 1982 issues.

8. Lawrence Livermore Laboratory. January 1983. "Status of Foreign Uranium Enrichment Activities." U.S. Department of Energy.

9. Lawrence Livermore Laboratory. April 1978, October 1980, and January 1983. "Status of Foreign Uranium Enrichment Activities." U.S. Department of Energy; author's estimates; historic U.S. exports and future estimates from U.S. Department of Energy, Enriching Operations Division, Oak Ridge Operations. June 1983. Reports 1475 and 1482.

10. For reports of lower secondary market sale prices, see, for example: *Nuclear Fuel*. August 15, 1983. "DOE Strike Against Discount SWUs May Be Late But Still Effective." p. 1.

11. International Atomic Energy Agency. June 1972. *The Structure and Content of Agreements between the Agency and States Required in Connection with the Treaty on the Non-Proliferation of Nuclear Weapons*. Vienna: IAEA, p. 9, Part II.

12. At the outset, Euratom included Belgium, the Netherlands, Luxembourg, France, Italy, and West Germany. In the late 1960s, the commission of Euratom, the European Economic Community, and the European Coal Commission were combined in the European Communities; the United Kingdom, Ireland, and Denmark became EC members in 1973, and Greece joined in 1980.

13. Intergovernmental Conference on the Common Market and Euratom. *Treaty Establishing the European Atomic Energy Community (Euratom) and other documents*. Brussels: The Secretariat of the Interim Committee for the Common Market and Euratom.

14. See Guidelines, Appendix A.

2 HISTORICAL BACKGROUND

The conditions surrounding nuclear fuel supply have deep historical roots. In addition to issues originating in the evolving industry structure, fuel supply has been influenced by the construction—and revision—of international political regimes for managing proliferation risks, and by the commercial ambitions of suppliers of uranium, fuel-cycle services, and reactors. Changes in domestic policies and political conditions in supplier and consumer countries also have had major effects, as has the struggle toward new forms of economic and political relationships between developing and developed countries. Finally, there have been fundamental changes in attitudes toward energy and its relationship to economic and political security.

In reviewing this history, it is convenient to talk in terms of four eras:

- *The emergence from military programs.* The late 1940s to about 1960: the initial development of reactor technology and fuel-cycle facilities under government sponsorship.
- *The surge of commercial and political development.* The years 1960 to about 1973: the beginnings of commercial development of nuclear power and the emergence of an international nonproliferation regime.
- *The period of conflict and instability.* Roughly 1974 to 1979: uncertainties, conflicts, and market failures in the context of heightened concerns about energy and security.
- *The post-INFCE era of market adjustment and policy accommodation.* The period following the close of the International Nuclear Fuel Cycle Evaluation: rapidly softening markets and a new pragmatism about nuclear fuel exports.

37

The boundaries between eras are not precise and the seeds of one era's problems (and of some of their solutions) can usually be found in preceding periods. Nonetheless, this simple breakdown does help in sorting out the events of the past four decades.

EMERGENCE FROM MILITARY PROGRAMS

In the United States, the era of commercial nuclear power began with the Atomic Energy Act of 1946; the first proposals for an international regime governing nuclear power were contained in the Baruch Plan of the same year. The U.S. act authorized civilian exploitation of nuclear-electric power, but with a federal monopoly on nuclear technologies and fuel. The Baruch Plan, presented to the United Nations in June of 1946, called for a similar arrangement internationally—that is, an International Atomic Development Authority managing all phases of the development and use of atomic energy, including nuclear fuel. The Baruch Plan eventually failed, and by the early 1950s independent nuclear research and power programs were going ahead in several non-nuclear-weapon states.

In December 1953, President Eisenhower delivered his "Atoms for Peace" speech before the U.N., calling for international cooperation in the development of nuclear power, including assistance in research and development and the provision, by the United States and other countries, of nuclear fuel and other materials. Implementation of the 1953 proposal required domestic U.S. legislation—the Atomic Energy Act of 1954—rescinding some of the secrecy provisions of the 1946 act and authorizing international cooperation.

This cooperation took the form of bilateral "Agreements for Cooperation" between the United States and foreign governments (twenty-two in 1955 alone). The agreements, which first emphasized research activities but eventually included power reactors and fuel, generally included safeguards and inspection provisions. The United States also reserved the right to approve plans for reprocessing fuel it had supplied, to approve retransfers to third countries, and to designate storage facilities for excess fissionable material (such as plutonium) or to purchase such excess material. Since all parties foresaw the eventual use of plutonium in nuclear power programs, these provisions were not seen as restricting its use for reactor fuel.

The creation of the International Atomic Energy Agency (IAEA), in 1957, presented opportunities to put safeguards and fuel supply under an international institutional umbrella. But delays in implementing such a regime—combined with the reluctance of some nations to put their nuclear futures in multilateral hands, and U.S. congressional reservations about a possible loss of influence—resulted in bilateral agreements continuing to dominate technology and fuel transfers for many years. Early expectations that the IAEA would function as an international fuel authority, with safeguards following flows of material, were never fulfilled.

One bilateral agreement that was to have major significance many years later was between the United States and Euratom. The Euratom agreement (which lapses in 1995) provided for supply of reactor fuel and special nuclear material for research; the supply is through the Euratom Community, which also took responsibility for safeguards. The initial agreements with individual member countries were subsequently allowed to lapse.

The 1950s also saw the beginning of power reactor development and deployment, first in the Soviet Union (a plant with five megawatts of electrical (MWe) output capacity in 1954) and then in the United Kingdom (four 50 MWe graphite-moderated natural uranium reactors in 1956). In the United States, development of the more complex pressurized-water reactor (PWR) for use in submarines led to the Shippingport nuclear power plant in 1957. The boiling-water reactor (BWR) was first utilized commercially at Dresden, Illinois, in 1959. Both reactor types made use of the low-enriched uranium producible in large quantities in the United States enrichment plants that had been constructed for weapons purposes in the 1940s and now had excess capacity.

During this period, uranium production was stimulated and sustained by the military procurement programs of the United States, the United Kingdom, and later France. Canadian production began in the early 1940s; a domestic U.S. industry was initiated with AEC encouragement in 1948; and production in Australia and South Africa began in the early 1950s, with purchases[1] by the United States and the United Kingdom through the Combined Development Agency. The United States imported uranium during this period from Canada, Australia, South Africa, the Belgian Congo (now Zaire), and Portugal. In all cases, production was encouraged by a variety of consumer and producer government incentives, including discovery rewards, guaranteed purchase prices, and tax concessions. Purchases of uranium by the United States, which were much larger than those of the United Kingdom, are shown in Figure 2-1. Data on U.S. purchases show that foreign procurements were consistently at higher prices than those in the United States between 1957 and 1967, when imports were phased out, though this difference was narrowed by other government subsidies and incentives in the United States.

Throughout this period, the United States played a dominant role, due to its general importance in the post-war world, its leadership in technology, and its monopoly position in nuclear fuel supply. From 1956 until mid-1961, the U.S. government—which had sole domestic control of U.S. uranium and enrichment—sold (or leased) enriched uranium for research and power uses to other countries under bilateral Agreements for Cooperation. At this time,

Figure 2-1.   U.S. AEC Purchases of Uranium.

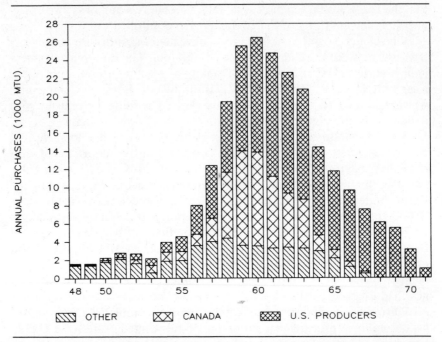

Sources: Data for 1956–71 from: U.S. Department of Energy. 1983. *Statistical Data of the Uranium Industry*. Grand Junction, Colorado: U.S. Department of Energy. GJO-100(83); pre-1956 deliveries are author's estimates based on U.S. and Canadian sources. It is not possible to separate imports in the "other" group, though the primary source in early years was the Belgian Congo (Zaire), followed later by South Africa and Australia.

the material was owned by the United States or the recipient government, not by private parties. In 1959, the U.S. Export-Import Bank began to finance sales of power reactors and fuel through loans and financial guarantees, thus signaling a national commercial interest in nuclear power trade.

THE SURGE OF COMMERCIAL AND POLITICAL DEVELOPMENT

The 1960s and early 1970s saw further development of the institutional and political framework, along with the first commercial investments in nuclear power. However, as discussed below, these developments did not deal entirely adequately with political problems, and conditions in the uranium industry were far from healthy. Thus, while the 1960s were an era of major progress, they also contained the seeds of problems that were to bring a crisis in the mid-1970s.

Growth in International Reactor Sales

In the 1960s, the first orders were placed for power reactor exports. By 1966, Canada had sold heavy-water reactors to Pakistan and India (a total capacity of 225 MWe), and the United Kingdom had sold Magnox reactors to Japan and Spain. Otherwise, the export market was dominated by the United States: Westinghouse sold eight PWRs (totalling 2,264 MWe) to seven countries, and General Electric sold eleven BWRs (2,369 MWe) to eight countries by the end of the decade. During this period, Canada was installing heavy-water reactors, and the United Kingdom and France were deploying gas-cooled, graphite-moderated reactors of their own design. In West Germany, AEG began to sell reactors under license from General Electric. In Japan, three companies began to develop light-water reactors for domestic use, under license from U.S. manufacturers.

Major changes came between 1967 and 1969 with the formation of Framatome in France, Kraftwerk Union (KSU) in Germany, and ASEA-Atom in Sweden. Direct government participation was involved in all but KWU. Framatome began to produce PWRs under Westinghouse license for domestic use. ASEA-Atom developed BWRs of its own design, and KWU began to develop its own PWRs. KWU secured its first PWR orders in 1969 (to the Netherlands), while Framatome did not make export sales until 1974, when it sold four PWRs (two each to Belgium and Iran). As will be seen below, the beginning of competition for international reactor sales put new strains on the nuclear fuel-supply system and on the nonproliferation regime.

Institutional Changes

In the 1950s, efforts to provide a political context for nuclear power development and nuclear fuel supply focused on bilateral agreements and regional integration in Western Europe. In the 1960s, increased attention was drawn to the need for a stronger global system. A comprehensive review of the role of the IAEA, as an alternative to bilateral agreements, was undertaken by the United States in 1961. While many recipient countries were happy with their bilaterals with the United States, the emergence of other suppliers, and the need to provide for universal safeguards in a nondiscriminatory way, provided incentives for bolstering the IAEA regime.

Beginning in 1964 (with intensive negotiations with India), efforts were made to shift bilateral safeguards agreements, old and new, to trilateral agreements including the IAEA. Simultaneously, the IAEA safeguards system was extended to include larger power reactors (1962) and reprocessing plants and other fuel-cycle facilities (1965). Enrichment plants were excepted. The IAEA system also was extended to include fissile material (e.g., plutonium) derived from material originally supplied under safeguards. The possible role of the IAEA as manager of fuel-cycle flows gave way to a role as administrator of safeguards, except in a few cases where the recipient country wanted the IAEA to act as the supply channel.

The major institutional achievement of this era was the Non-Proliferation Treaty (NPT) which was opened for signature in 1968 and which took effect in 1970. As discussed in Chapter 1, the treaty represents an international compromise between weapon and non-weapon states, involving pledges of peaceful nuclear cooperation in exchange for agreement not to develop nuclear weapons and to accept safeguards on *all* peaceful nuclear power activities. The interpretation and implementation of these provisions has been a continuing and controversial process, with significant consequences for trade in nuclear fuel. The problem of an independent Euratom safeguards system, originally accomplished through painful political accommodation in Europe, was resolved by an understanding that the Euratom non-nuclear-weapon states would negotiate a safeguards agreement collectively with the IAEA. The fact that Euratom involves both weapon and non-weapon countries, while the NPT makes a fundamental distinction between them, posed a special problem at the time, and more recently, as noted below. However, from the standpoint of nuclear power development and the stability of fuel-supply arrangements, this era was one of increasing optimism about the possibility of separating nuclear power from nuclear weapons.

Developments in the Fuel Cycle

Until the early 1970s nuclear fuel supply was primarily an issue of enrichment services, and the history of enrichment was mostly one of U.S. policy initiatives. From a consumer perspective, uranium supply was not a problem; the production capacities built up in the 1950s were far larger than commercial demand. Moreover, as the principal source of enrichment, the United States often provided the uranium from its own stockpiles, which were increasing due to domestic purchase programs. Outside the United States, the uranium industry picture was one of severe depression, in the trough between military use and buildup of civilian nuclear power.

Uranium. In 1959, the United States, foreseeing a saturation of weapons needs, announced that it would no longer make foreign purchases of uranium; most existing purchase contracts were to expire by the early 1960s. The result, especially in Canada, was the near collapse of the uranium industry. Canadian production dropped from more than 12,000 STU_3O_8 in 1959 to about 3,000 STU_3O_8 in 1966. Even this level was sustained only through a government stretch-out program, a transfer of contracts to low-cost producers, and the buildup of a government stockpile. Only four out of twenty-eight producers remained active.

In Australia, the impact was not as great, due to the relatively low level of production and the high degree of government participation and stock-pile building. In South Africa, the impact of reductions in exports was small since most uranium production was a byproduct of gold mining; the uranium actually produced after 1960 (about half the peak rate) was stockpiled. In 1967, the South African government legislated private ownership of uranium and transferred its calcining facility to the Nuclear Fuels Corporation of South Africa, Ltd. (NUFCOR), which acts as the uranium marketing agent for the gold-mining companies and is subject only to the export controls of the South African Atomic Energy Board.

In the United States, government stimulation of the uranium industry ended with a moratorium on new contracts in 1958. From 1962 to 1966 the AEC carried out a maintenance program in which there was an annual 500 STU_3O_8 (385 MTU) limit per property and a fixed price of $8 per pound. The program sustained the industrial base while limiting the further growth of what was already a large stockpile (about 50,000 STU_3O_8 (38,460 MTU), excluding military stocks). However, the reduction in government demand—from a high of 17,600 STU_3O_8 (13,540 MTU) in 1961 to 10,200 STU_3O_8 (7,850 MTU) in 1966—resulted in considerable contraction in the domestic industry and reduced exploration. When expected power-plant demand failed to materialize, the AEC began to stretch out its contracts; by 1970 prices had dropped to an average of $6 per pound. This price history was shown in Table 1-7. Prices outside the United States fell even lower. Prices of uranium in nominal terms had fallen since the mid-1950s, but the fall in *real* prices was even more severe.

In 1964, the Atomic Energy Act was amended to allow private ownership of nuclear fuels. In 1966, a year which also saw the first big surge in domestic reactor orders (twenty reactors with a total capacity of 16,400 MWe), the United States instituted an embargo on the import of foreign uranium for enrichment for use in domestic reactors. This move isolated foreign producers from the first surge of U.S. demand (only three reactors were ordered outside the United States in 1966). The first private purchases of uranium in the United States began in 1967 and rose rapidly to 12,700 STU_3O_8 (9,770 MTU) in 1971, when the AEC ended its purchase program. As a result, uranium demand in the United States was kept relatively constant during the transition from military to civilian use.

The net result of these events was severe depression in uranium industries outside the United States and a static, but not especially disruptive, environment in the United States. This pattern did not have immediate consequence for commercial nuclear power development; however, it was to have profound consequences later.

Enrichment.   The effects of fuel-cycle developments on nuclear power came most immediately from changes in U.S. enrichment policy. The 1964 Atomic Energy Act changed the terms of enrichment availability for domestic and foreign customers. The new policy allowed toll enrichment of uranium procured abroad, though the AEC would also sell uranium from U.S. stocks if requested. Whereas previous procedure had been to provide whatever amounts of enriched uranium might be desired by bilateral partners, the new policy was to provide material under long-term contracts. The purpose of these new contracts, beginning in 1968, was to allow longer term planning by the builders of an expected wave of new power plants and by the AEC in its enrichment operations.

The revision in U.S. enrichment contracting represented only a small change in the role of the United States in the international fuel-supply system. However, it came at a time when the international commercial context was changing. Other industrialized countries were beginning to enter international reactor markets, and they very likely saw the U.S. monopoly of enrichment as putting them at a commercial disadvantage.

In an effort to deal with the potential spread of enrichment technology—and especially the centrifuge, which was regarded as more proliferation-sensitive than gaseous diffusion—the United States, in 1971, proposed a multilateral approach to new enrichment capacity. However, the terms of this proposal seem only to have reinforced suspicions abroad that U.S. security concerns were really a cover for commercial ambitions: not only was the proposed technology transfer apparently one-sided (with the prospect that the United States would learn more about European technology than vice versa), but the United States explicitly required protection of its commercial interests in supplying components for whatever plant might be built and insisted on continued access to all markets for U.S.-supplied enrichment services.

The "privatization" of nuclear fuels—part of an overall effort by the AEC and the Nixon administrations to put all of nuclear power on a commercial footing—implied that fuel supply could become tied to private commercial ambitions in the United States as well as to governmental international security interests. The U.S. dominance of reactor orders abroad only increased such concerns. While commercial motives may not have been dominant in U.S. decisions, attitudes abroad clearly reflected a growing concern about U.S. commercial dominance and revealed the difficulty of distinguishing between commercial and international security motivations in the new atmosphere of international competition.

One result of these concerns was increased interest in European enrichment projects. By 1968, Foratom, the European nuclear industry organization, had begun plans for ventures that would provide increased autonomy. In 1970, Germany, the Netherlands, and the United Kingdom established Urenco, an enrichment venture based on centrifuge technology. In 1972, the Eurodif enrichment consortium was chartered, using French diffusion technology. Also in 1970, European utilities and reactor companies began negotiations with the Soviet Union's Techsnabexport for considerable quantities of enrichment services to be delivered between 1974 and 1990. West Germany was the largest purchaser, though others included Sweden, Spain, France, Belgium, Finland, Italy, Austria, and the United Kingdom. These contracts with the Soviet Union served to decrease dependence on the United States during the few years remaining before Urenco, Eurodif, and other ventures reached full output.

THE PERIOD OF CONFLICT AND INSTABILITY

By the early 1970s, a number of processes were under way that ultimately would alter perceptions of nuclear fuel security and affect the viability of the nuclear option itself. There were changes in U.S. policy regarding enrichment, including new efforts to transfer enrichment to the private sector and a change to Long-Term Fixed Commitment contracts. In 1974, the United States closed its books to further enrichment orders. At the same time, the uranium industry was on its way from a buyer's to a seller's market, in part because of a rise in demand, but also because of the demand induced by changes in U.S. enrichment-contracting policy. Additional disruptions were induced by the formation of a uranium cartel, the unexpected delay in Australia's emergence as a prospective supplier, massive sales by Westinghouse of uranium for which it did not have contracts with primary producers, changes in Canada's rules for holding domestic reserve margins, France's withdrawal from sales of uranium, and a U.S. import embargo.

Concurrently with these changes, competition for reactor orders was increasing with the entry of European vendors, and the focus of sales efforts was shifting to the developing countries. This competition, combined with the drive for nuclear autarky in Western Europe, served to accelerate technological change and increase the pace of commitments to plutonium fuels, breeder reactors, and indigenous enrichment and reprocessing plants. And in the midst of all this came the Indian nuclear explosive test. The Indian explosion, coupled with plans for transfers of proliferation-sensitive technologies to other less developed countries, raised fundamental questions about the effectiveness of the existing nonproliferation regime, and led to retroactive as well as prospective changes in the political conditions for fuel exports from the United States and Canada.

All these events occurred in the space of about five years and were mutually reinforcing. The net result was a sharp decline in the perceived security of nuclear fuel supply and a sharp rise in the price of uranium. In the following subsections, we look at these events in more detail, focusing on developments in enrichment and uranium markets and in the national policies that determine the conditions of nuclear fuel trade.

Enrichment

Because of the dominant role of the United States in enrichment, its domestic policies regarding ownership, operation, and contract terms could not avoid affecting nuclear fuel supply and nuclear development. The policy changes of 1971 to 1975 had a particularly large impact, the most important events being the introduction of Long-Term Fixed Commitment contracts and closing of the U.S. enrichment order books, and the associated changes in plans for uranium stockpiles and enrichment tails assays. During this same period, the U.S. enrichment dominance also helped precipitate European commitments to new sources of enrichment supply.

At the end of 1972, the United States had entered into Requirements contracts for 107,000 MWe of nuclear capacity, 25,335 MWe of which was for foreign customers. In January 1973, the AEC announced its intention to offer only Long-Term Fixed Commitment (LTFC) contracts, and in September 1973, the commission began to accept LTFC orders. Unlike the flexible Requirements contracts, the LTFC contracts required that firm commitments for enrichment of initial cores be made eight years in advance, and that reloads (usually ten-years' worth) be committed soon thereafter. Moreover, the AEC announced that contracts would be issued only for reactors requiring enrichment of initial cores before July 1982. Heavy penalties would result from deferral or cancellation of LTFC contracts.

By July 1974, the Energy Research and Development Administration (ERDA), which inherited the AEC enrichment functions, had executed LTFC contracts for 166,000 MWe of reactor capacity, 42,000 MWe of which were with foreign buyers. Unfilled requests totaled 91,000 MWe (75,000 MWe foreign). Additional firm contracts (above nominal enrichment capability, if the enrichment plants were operated at 0.3% tails assay) were written, but 44,000 MWe of reactor capacity (mostly that of foreign customers) remained unsatisfied. Contracts were offered for this capacity, conditional on the United States proceeding with plutonium recycle; twenty-seven of these conditional contracts were written. To deal with the uncertainties imposed on these customers, President Nixon announced in August that the United States would "in any event" fulfill the conditional contracts.[a]

The effects of the LTFC contracts and the closing of the U.S. order books were significant. Not only did the contracts seem to encourage commitments to large numbers of reactors worldwide (an effect consistent with the large number of reactor orders in 1973 and 1974), but the substantial long-term commitments put considerable pressure on an already tight uranium market (as discussed below). These effects, combined with a decline in growth of electricity demand, environmental opposition, and other factors, led to intense pressure in 1975 for readjustment of the LTFC contract arrangements. ERDA responded in mid-1975 with a "once and for all" open season—a period in which delivery, and thus reactor schedules, could be slipped. However, first cores were not allowed to slip beyond 1985, and a fraction of the natural uranuim feed had to be delivered on the original schedule.[b] As of mid-1976, foreign capacity covered by LTFC contracts stood at 78,834 MWe.

Even more important than the pressure on reactor procurement and fuel-cycle activities was the effect on the *perceptions* of foreign consumers and suppliers. These events damaged the confidence of consumers in the reliability of the United States as a long-term supplier, and some foreign reactor suppliers initially saw the U.S. move as an effort to induce commitments for new reactors that would be purchased from U.S. vendors. The U.S. move could also have been seen as a preemptive effort to tie up enrichment demand before new ventures abroad were in a position to write contracts.

In fact, the changes in U.S. contracting policies are probably best understood as resulting from domestic efforts to shift enrichment to the private sector, begun in the late 1960s, and the desire to make long-term enrichment planning more secure in anticipation of a wave of new reactor orders expected in the late 1970s and early 1980s. While the reasons for privatization were probably more domestic than international—the original intention had been to make nuclear power a private endeavor and there was pressure to cut the size of the AEC budget—there appears to have been suspicion abroad that the United States was beginning to convert its traditional "promotion in the name of international security" into a drive for commercial dominance in an increasingly competitive world market. Whatever the facts, these perceptions could only add to the growing uncertainty about future availability of enrichment supplies from the United States and accelerate commitments to supply from the Soviet Union and from the new European enrichment ventures.

[a]Of the twenty-seven, eighteen holders of conditional contracts later terminated, some because they had been assigned a LTFC contract by an earlier purchaser. In 1977, seven of the remaining contract holders terminated and two (both in South Korea) converted to LTFC contracts.

[b]For contracts originally beginning in fiscal year 1976–78, 100 percent of feed would be required; for 1978–80, 50 percent; and 1981–83, 25 percent.

Uranium

Until 1973, the worldwide uranium industry remained weak. There had been a surge of reactor orders by U.S. utilities in the early 1970s, but it does not seem to have had much effect even on U.S. spot prices, and in any event, the rest of the world was excluded from the U.S. market by the continuing embargo on use of foreign uranium. Utilities and consumer governments generally believed that uranium would be available at low prices; there was little interest in long-term contracts, buyers preferring the spot market where prices were falling in real terms in mid-1973.

The U.S. government also provided an additional demoralizing shock: just as reactor orders were picking up in the early 1970s, the AEC proposed in late 1971 to dispose of 50,000 STU_3O_8 (38,460 MTU) from the U.S. stockpile. To reduce, or at least delay, the impact on the uranium market, the AEC devised a "split-tails" contracting arrangement, announced in April 1972. Utilities would deliver uranium as if the enrichment plants were to operate at 0.20 percent tails assay. However, the plants would actually operate at 0.25 percent tails assay with the resulting requirement for additional uranium to be met from the AEC stockpile. This scheme would delay the effect on the uranium market compared to what might be experienced with a direct sale. However, it did mean that U_3O_8 demand would be about 20 percent lower than otherwise; it also showed that changes in enrichment contracting could suddenly alter uranium market conditions. Because the split-tails policy affected foreign as well as domestic enrichment buyers, the U.S. move undermined demand in the non-U.S. uranium market. In effect, the United States began to export uranium at prices (collected through enrichment fees) that were high relative to those in the international uranium market.

With a depressed uranium market in most uranium-producing countries, the atmosphere was created for government intervention, protectionist measures, and cartel formation. In the spring of 1971, a series of meetings began that were to culminate in the "Club" or cartel of producers, which was active from 1972 to 1974. The government of Canada was apparently responsible for the first initiative through discussion with Australian officials about uranium marketing strategy. In May 1971, Canadian Resources Minister Greene was quoted[2] as saying that it was in the interest of Canada and Australia to work together on pricing, and that it was likely that agreement would bring in the large private producers like Rio Tinto Zinc. In the fall of 1971, following a June visit by Canadian officials to the United States to argue for relief from the import embargo, the AEC restated the U.S. intention to continue its import embargo until the late 1970s.

By early 1972, there were reports of a meeting in Paris of representatives from France, Canada, Australia, and South Africa intended to "put some order into the international uranium market . . . to coordinate uranium production and marketing policies.''[3] Subsequent meetings were held in Paris, Johannesburg, and Cannes through at least February 1974. Cartel documents released later[4] showed the development of a plan to allocate market shares for two periods, 1972–77 and 1978–80, and to establish minimum prices and bidding priorities for sales. The domestic markets of France, South Africa, Australia, Canada, and the United States were excluded. Minimum prices were set to rise from around $5.50 to near $8.00 in the period 1972 to 1978. There is no question that there was a uranium cartel. The much more difficult question is whether the cartel actually had a significant effect on prices. In the United States, this question has been the subject of extensive litigation. Whatever the merits of the associated arguments, the origin of the cartel is plausibly viewed as a symptom of the depressed market conditions of the earlier era.

Conditions in the uranium market did ultimately tighten, and prices rose far beyond the expectations of the cartel organizers. While it is possible that the cartel assisted the price increases, it is the author's view that there were more than adequate causes for prices to rise as they did, without the cartel. The cartel may have affected *how* things happened, but it seems unlikely that it fundamentally influenced *what* happened. This view does not settle the legal questions. However, it is adequate to our historical and economic assessments.

Sufficient causes for market tightening and price increases were suggested earlier. U.S. enrichment policy—and the new commitments to enrichment supply from Eurodif, Urenco, and the Soviet Union—had a strong effect. The introduction of LTFC contracts and the new foreign enrichment commitments produced a surge in uranium demand in late 1973 and early 1974. The fact that these enrichment commitments implied a complementary need for long-term uranium feed converted many utilities from their historic spot-purchase strategy to a new emphasis on long-term contracts. This change was initially more striking outside the United States than among U.S. utilities.

Changes in uranium supplier government policies had similar, but smaller, effects: in 1972 a newly elected Labor government put a lid on Australian exports; Canada (in September 1974) adopted a domestic reserve policy which required that a fraction of reserves be set aside for domestic use; and France (in April 1974) withdrew from the supply of uranium due to its new commitment to a much-expanded reactor program and fears about continued supply from unstable African nations. Finally, there was the Westinghouse abrogation of uranium supply contracts that was announced in mid-1975. The point when Westinghouse's short position became generally known has been an object of contention in litigation. There is evidence that some industry analysts[5] knew of disparities between Westinghouse's supply and commitment positions in 1973 (though it is not clear what might have been made of this information). It was not until late 1974 that public indications appeared in the trade press. The chronology of these events is shown in Table 2-1.

Of these events, the largest effects appear to have come from the large commitments to new reactor capacity, partly due to the introduction of Long-Term Fixed Commitment contracts and other long-term enrichment commitments, and (in the United States) from the Westinghouse abrogations. The effect of the new enrichment contracts was twofold: utilities were forced to take a longer term view of procurement, and the new demand represented a sizable increase over previous expectations. The first cores (initial fuel loads) alone under the new contracts required procurement by foreign customers of an extra 26,900 MTU (at 0.20% enrichment tails assay)[6] before about 1980, compared to deliveries under Requirements contracts of about 15,380 MTU. Annual requirements for reactor reloads were to be increased comparably after 1980. The increase in domestic U.S. delivery commitments was much smaller.

The net result of all this was a large shift upward in demand, occurring at the time when Australia had indicated its unwillingness to enter the market, and Canadian and French supplies were being reduced. At the end of 1973, outstanding U.S. utility invitations for bids stood at 40,000 STU$_3$O$_8$ (30,770 MTU).[7] As shown in Table 1-4, spot prices began to rise in 1973, probably more slowly than they might have under these circumstances since Westinghouse's short position was still not widely known, thus isolating suppliers from the growth in demand.

Prices continued to rise during 1974, doubling by the end of the year. During 1974, U.S. producers and agents sold 17,600 STU$_3$O$_8$ (13,540 MTU) to domestic buyers and 5,200 STU$_3$O$_8$ (4,000 MTU) to foreign purchasers, who were appearing in the United States for large quantities for the first time (the previous year saw foreign sales of only 500 STU$_3$O$_8$). Moreover, in 1974 the AEC announced that the U.S. ban on domestic use of foreign uranium would be lifted beginning in 1979; by the end of the year, U.S. firms had contracted abroad for 25,400 MTU. (Previous U.S. purchases abroad were reported by the AEC as being only 5,400 to 6,150 MTU total.[8]) The procurement activities of U.S. utilities abroad increased the pressure on supplies available to foreign utilities.

In the resulting seller's market, the first "market price" contracts were written (the Canadian Rio Algom contracts with Duke Power and TVA). Such market price contracts were symptomatic of the great uncertainty being experienced by both sellers and buyers. Other changes also began to appear. Increasingly, utilities moved to arrange procurement directly: by the end of 1974, 68 percent of total U.S. forward-delivery commitments had been ar-

Table 2–1.   Events Affecting the International Uranium Market.

1958		United States announces moratorium on new import contracts.
1964		United States announces prohibition of enrichment of foreign uranium for domestic use, effective 1966.
1971		
	October	AEC proposes to dispose of its stockpile beginning in 1974.
		AEC announces postponement of lifting of import restrictions until late 1970s.
1972		
	February	Uranium producers meet in Paris to discuss "orderly marketing."
	March	United States reaffirms continuation of import embargo until late 1970s.
	April	AEC announces split-tails enrichment policy to dispose of 50,000 short-ton stockpile of uranium.
	May	Johannesburg meeting of uranium producers.
	July	Canada imposes new controls on price and quantity of uranium exports.
	December	Australia elects Labor government; hold placed on new export contracts.
		United States announces consideration of change to long-term enrichment contracts.
1973		
		Increase in gold price indicates future reduction in South African uranium output.
	January	U.S. AEC announces switch to Long-Term Fixed Commitment enrichment contracts.
	September	United States begins to accept LTFC contracts; order books to close in June 1974.
	October	TVA request for bids on 33,000 MTU.
		Gabon establishes diplomatic relations with Soviet Union; calls for creation of an organization of uranium-exporting nations; declares national control of uranium resources.
	November	Arab oil embargo increases consumer concerns about supply security and suggests extent of producer market power.
		United States proposes end to restrictions on imports for domestic use.
1974		
	April	Military coup in Niger. France's uranium marketing agency, Uranex, withdraws from new uranium export commitments.
	May	India explodes nuclear device.
	June	United States closes enrichment order books.
	Fall	Persistent industry rumors of Westinghouse short sales.

Table 2–1. *(continued)*

1975	September	Canada announces domestic reserve set-aside and requires new safeguards conditions on existing and future export contracts.
		New Australian uranium policy featuring government involvement in all stages of industry and new export conditions.
	October	United States announces end to import restrictions beginning in 1977; U.S. buyers enter international market.
	July	Westinghouse asks for price renegotiation of its existing sales commitments; not enough uranium secured at reasonable prices.
	September	Westinghouse announces shortfall of more than 25,000 MTU; claim of "commercial impracticability."
	December	New Australian government; uranium to be returned to private sector activity.

ranged directly by utilities. The remainder were arranged by reactor vendors or other agents. Late in 1974, rumors about the Westinghouse situation began to surface. So also did suggestions by ERDA that it would have to raise the tails assay on which uranium delivery requirements were based (a result of a revised stockpile policy), thus increasing the amount of uranium that United States and foreign utilities would have to deliver to the enrichment plants.

Uranium prices doubled again in 1975 in this volatile atmosphere. ERDA's open season on Long-Term Fixed Commitment contracts gave an opportunity to reduce demand pressure for deliveries in the late 1970s. But Westinghouse made its first disclosure of its short position in July, and in September claimed "commercial impracticability" and declined to deliver on contractual responsibilities.[9] This claim was initially based on the OPEC price increases and embargo but later was changed to allegation of uranium cartel price manipulation. The Westinghouse contracts were virtually all with U.S. utilities; abroad, only Sweden's State Power Board was involved. The result was another scramble for new contracts and a further bidding-up of prices. The main effect appears to have been on the domestic U.S. market: U.S. purchasers contracted for 12,460 MTU from domestic suppliers and 3,380 MTU from foreign sources in 1975; in contrast, U.S. producers sold only 690 MTU to foreign buyers.[10]

In 1976, there were further changes in the U.S. uranium market as consumers and producers responded to growing uncertainties with a wave of vertical integration. Instead of contracts for deliveries, producers began to propose joint ventures with long-term financing arranged by utilities. This process had actually started during 1975, but the major impact on procurements was delayed until 1976 when a record 92,900 STU$_3$O$_8$ (71,460 MTU) were contracted between domestic U.S. producers and consumers. Of this quantity, some 47 percent was from primary sources in which purchasers had a direct involvement.[11]

A similar picture was emerging outside the United States. However, the energy security interests of countries like West Germany and Japan, and the risks of making investments abroad, led to relatively high levels of government involvement, either directly or through financing, guarantees, or other subsidies. Government backing for foreign uranium ventures also reduces the risk of supply interruption by host country governments, particularly where the latter is in joint venture with the foreign entity.

The development of these patterns of integration appeared to have stabilized the uranium market. Prices stopped their upward spiral and indeed fell slightly in constant dollars starting in mid-1976. The new procurement level in the United States fell back to 12,000 STU_3O_8 (8,600 MTU) in 1977,[12] a figure comparable to annual production levels.

The uranium and enrichment market stabilization which began in 1976 was not the end of the story on fuel supply concerns, however: just as market problems appeared to be resolved, a series of political events further disrupted nuclear fuel-supply arrangements.

The Great Policy Shifts

The new difficulties arose out of an increasing politicization of nuclear exports due to proliferation concerns. The precipitating events were the Indian explosion (in 1974), and trade deals involving the transfer of reprocessing and enrichment technology (in 1975 and 1976). The planned technology transfers were from France and Germany—as new suppliers—to Pakistan, South Africa, South Korea, and Brazil. These events reflected a change in the international balance of commercial and political influence in nuclear matters. The entry of new suppliers reduced the leverage of traditional nonproliferation leaders, like the United States, and this happened at a time when technological change and international technology tranfers were bringing into question the capabilities of the nonproliferation regime negotiated during the preceding decade. The consequences for assurance of fuel supply were profound, since nuclear fuel supply was the primary form of direct leverage over the nuclear activities of other countries retained by traditional suppliers.

The Indian Explosion.   India's first nuclear device was exploded on May 18, 1974. It had been fabricated using plutonium produced in the "Cirus" research reactor supplied by Canada under a "peaceful uses" agreement. The event was a startling reminder to supplier nations that bilateral agreements and other arrangements could not prohibit misuse of transferred technologies and materials. It also revealed an important ambiguity in bilateral and multilateral agreements. Even prior to India's nuclear detonation, disagreement arose between Canada and India as to whether all nuclear explosions were precluded by the peaceful use provision. India maintained, and does to this day, that peaceful nuclear explosives were not prohibited.

The Indian explosion had its most immediate effect on Canada, which considered the Indian explosion to be in violation of their bilateral agreements. Canada stopped all nuclear cooperation with India, terminating fuel supplies to the U.S.-exported Tarapur power reactor. Canadian authorities also initiated reconsideration of export policies, which had not been a major political issue in Canada until the Indian explosion. The Indian explosion thus propelled Canada to the forefront of the proliferation policy issue; the result was new conditions on access to nuclear fuel and disruption of supply.

Technology Transfer and Reactor Sales.   The efficacy of the existing non-proliferation regime was also called into question by the effects of competitive pressures for exports and the use of sales and technology transfers as instruments of national policy abroad. In 1974, the Nixon administration offered nuclear reactors and fuel to Egypt and Israel. It was argued by the administration that American nuclear technology should be used as a general tool of diplomacy, and if the United States did not make the sales, other, less responsible, suppliers would. However, there were complaints in the U.S. Congress and elsewhere that the pursuit of diplomatic objectives through nuclear sales might ultimately increase the chance of nuclear conflict in the Middle East. There were also questions about whether such visible sales to non-NPT signatories were in the best interests of U.S. policy. This use of nuclear sales for general diplomatic and commercial purposes may have weakened U.S. nonproliferation leadership in a key period of international change. The arguments used to justify the U.S. offer to Egypt and Israel were later echoed by other nuclear technology suppliers entering the export market.

The political context of international nuclear trade was further complicated by the multibillion-dollar sales by France and West Germany to Brazil, Iran, South Korea, Pakistan, and South Africa. The nuclear export market was becoming extremely competitive, with not only power reactors but fuel-cycle equipment becoming involved. The political leverage of any one exporter was thus reduced at the very time when sensitive nuclear technologies were spreading over the globe. While U.S. policy was to prohibit the export of reprocessing and enrichment plants, France and Germany were offering both to foreign customers, giving them a competitive advantage over American manufacturers.

In June 1975, Brazil and West Germany signed a contract calling for a large transfer of all stages of fuel-cycle technology. The deal involved the possible construction of eight power reactors and certain privileges for Germany in exploiting Brazilian uranium resources: included were an industrial-scale enrichment plant, a fuel fabrication facility, and a pilot-scale reprocessing plant. Though Brazil agreed to safeguards stricter than those of the IAEA, there was concern that these controls could not be sustained after transfer of the facilities. Brazil had not signed the NPT and claimed that the safeguards agreement in no way prejudiced its access to "nuclear devices" in the future.

France also became involved in commercial activities involving "proliferation-sensitive" technology. In 1975, France entered negotiations for the sale of a reprocessing plant to South Korea, and in 1976 agreed to sell a reprocessing plant to Pakistan. In each case, the deal involved a safeguards package to be negotiated between France, the buyer, and the IAEA, and promises not to use equipment or materials for the manufacture of explosive devices. However, even these arrangements raised great concern in the United States and other countries. South Korea suspended negotiations in early 1976; the French supply arrangement with Pakistan was subsequently suspended.

There was sharp competition for reactor sales in other countries as well. Iran made major purchases from Germany in 1976 and France in 1977, all against U.S. competition. Germany did not promise reprocessing or enrichment technology to Iran, though it was reported that the two countries were discussing an option for a reprocessing plant.[13] Also in 1976, a French consortium signed its first contract with South Africa. In this case, General Electric had submitted the lowest bids and expected to get the sale. One possible reason for the choice of the French offer was the difficulty G.E. had in guaranteeing the supply of components and fuel, given the complexities of the Nuclear Regulatory Commission (NRC) procedure and the possibility of a confrontation over South African racial policies. There were also reports that France had coupled the reactor deal to conventional military weapons sales to South Africa,[14] a conjecture officially denied by the French government.

Whatever the details, it is clear that over the period of 1974 to 1976 fundamental changes had taken place in international trade in nuclear equipment. Strong competition was established, altering the balance of nuclear influence in the world. Moreover, the transfer of fuel processing technologies became an important element in the bargaining, and differing national policies on technology transfer further modified the former pattern of commercial relationships.

Canadian Policy Changes.   Canada, whose exports had been used by India to produce a nuclear explosive, responded first with a series of changes in export policies. In December 1974, Canada called for a renegotiation of existing agreements and the retroactive and prospective imposition of new nonproliferation conditions on all uranium contracts with other countries. While Canada was able to modify agreements with some countries (for example, Argentina), progress with Switzerland, Japan, and the Euratom countries proved difficult. The renegotiation period ended in December 1975. Two subsequent six-month extensions were then granted, during which time there was increasing conflict, first over French objections to verification of IAEA safeguards, and then over escalating Canadian demands for the right to inspect facilities using Canadian uranium.

At the end of 1976, Canada stopped exports to Japan and the Euratom nations, pending agreement to its new conditions. Under the new regulations, new contracts, or contracts pursuant to existing agreements, would be approved only if the consumer country accepted the NPT or agreed to equivalent safeguards on its entire peaceful nuclear program, a provision commonly referred to as "full-scope" safeguards. Canada also required a prior approval condition on reprocessing and retransfers to third parties, a pledge not to develop "peaceful" nuclear explosives, and implementation of Euratom-IAEA agreements on the latter's safeguards role.

Euratom posed a special problem for the Canadian initiative. As discussed above, the formation of Euratom had involved major political accommodations within Europe, including agreement to free transfer of material within the Euratom Community and the establishment of an internal safeguards system. The European Community members thus contended that Canada had to deal with Euratom as a whole, despite the fact that most uranium contracts with Canada had not been initiated through Euratom. Canada's position was that material could not be transferred to France, which had not signed the NPT or executed a trilateral safeguards agreement with Euratom and the IAEA (as had the other members of Euratom).

The Canadian interruption of uranium shipments was relaxed by early 1978, with a temporary remission of disagreements aided by events outside Canada. Policies were changing in the United States, as discussed below, and the International Fuel Cycle Evaluation (INFCE) was begun. INFCE provided an opportunity to deal with the Canada-Euratom dispute by postponing resolution of the renegotiation issue until the year following the completion of INFCE. During the interim period, Canada suspended its original demands for control over reprocessing, enrichment, and retransfer with only "prior consultation" requested. The interim agreement, developed in December 1977, was made without prejudice as to the outcome of ultimate negotiations: to grant Canada a suspension of its requirements would have implied Euratom recognition of Canada's power to impose such conditions.

The interim agreement thus represented a suspension of the sensitive prior-approval issue (a position weaker than that in the U.S. legislation discussed below). In exchange, however, Canada did achieve the implementation of trilateral safeguards agreements with the non-weapons states in Euratom and the IAEA, and with the IAEA and the United Kingdom. France, the principal stumbling block in the Euratom negotiations, was to continue negotiation of a separate agreement with the IAEA.

In addition, there was the negotiation (in December 1977) of an interim Canadian-U.S. agreement on "double-labeling" (i.e., the imposition of separate safeguards systems by two countries). Under the agreement, the United States was committed to consult with Canada concerning imposition of safeguards conditions prior to releasing Canadian-origin material (e.g., following enrichment in the U.S.). This arrangement provided the key to resolution of a disagreement between Canada and Japan. Japan renegotiated its bilateral agreement with Canada to reflect the new Canadian conditions in January 1978. The new agreement provided for Canadian approval of safeguards on reprocessing, enrichment, storage, and retransfer.

A factor that increasingly affected incentives for Canadian accommodation was the economic significance of uranium exports to the nation and to the uranium industry; both remembered the hard economic times that had only recently given way to rising uranium sales at increasingly high prices. As early as March 1977, news reports indicated mounting pressure from the uranium industry for resolution of the safeguards deadlock. The embargo had tied up contracts worth more than $300 million. Late in 1977, the Canadian Trade Minister was quoted as saying that Canada was waking to the "commercial realities" of its safeguards policy.[15]

U.S. Policy Changes.   While the Indian explosion stimulated an early response in Canada—due to the direct involvement of Canadian equipment—the effect on U.S. policy was slower to develop. The United States first responded to the German and French technology transfers by attempting to intervene directly with the countries involved or indirectly through the London Suppliers' Group; basic shifts in nonproliferation policy came later. In part, this delay was due to the fact that nonproliferation policy was complicated by a growing pluralism and ambiguity in the policy formulation process.

The Energy Research and Development Reorganization Act of 1974 began to open up what had been a monolithic nuclear policy process within the AEC and the Joint Committee on Atomic Energy. Licensing of exports was assigned to an autonomous Nuclear Regulatory Commission (NRC) that began to play a day-to-day, independent role in interpreting nonproliferation conditions. Such a system would have been relatively stable in an era with little change in international nuclear problems and little need for change in U.S. policy. However, with rapidly changing international conditions, the NRC was put in the uncomfortable position of having, in its routine licensing decisions, to play an important foreign policy role.

In March 1975, the NRC began a policy of closer scrutiny by the commissioners themselves of potentially sensitive exports; this change in procedures delayed licenses and was widely interpreted abroad (and still is) as an export ban. In West Germany, the Research and Technology Minister stated that the "export ban underlines the need to become as independent as possible from foreign energy sources."[16] That even small changes in procedures could raise such concerns revealed a growing uncertainty about the reliability of U.S. supply and increasing sensitivity to the security of nuclear fuel supply generally.

The NRC also had to respond to the effects of the Indian explosion and to a growing awareness of the difficulty of making reprocessing amenable to traditional safeguards measures. The issue was brought to the fore early in 1976 with consideration of an export license for fuel for India's Tarapur reactors. For the first time, nuclear opposition groups (the Natural Resources Defense Council, the Sierra Club, and the Union of Concerned Scientists) sought to intervene in opposition to the granting of a fuel export license, citing the lack of Indian acceptance of the NPT or adherence to full-scope safeguards. The announced purpose of the intervention was to force alteration of U.S. policy—an issue that went beyond the functions of the NRC. The license dispute was clearly focused on the reprocessing of plutonium, with consideration given to U.S. buy-back of spent fuel. When the license was finally issued at mid-year, it was with a dissent by Commissioner Gilinsky contending that safeguards on plutonium resulting from the irradiation of U.S.-supplied fuel could not be considered adequate.[17]

While the NRC could only respond to growing proliferation concerns in its application of existing law to specific exports, 1976 also saw the beginnings of the legislative revolution that would ultimately culminate in the Nuclear Nonproliferation Act of 1978. Activities in the 94th Congress appear to have been motivated by the sensitive technology exports undertaken by West Germany and France. Attention centered not only on conditions for U.S. exports but also on the exercise of U.S. leverage on the policies of other suppliers; it also focused on the roles to be played in the policy process by a host of governmental entities,[a] with increasing importance attached to the role of a Congress experiencing a resurgence of interest and power in foreign policy. However, action in Congress was itself inhibited by splits between committees: the Joint Committee had to yield its monopoly on domestic nuclear policy to the Government Operations Committee and the Foreign Relations Committee in the Senate and to the International Relations Committee in the House only when export issues were considered. The result— until the demise of the Joint Committee in 1977—was difficulty in passing new legislation.

[a]Including the NRC, ERDA, the State Department, the Department of Commerce, the Arms Control and Disarmament Agency, and the National Security Council.

Legislation *considered* by the 94th Congress in the summer of 1976 included virtually all of the provisions of the Nonproliferation Act of 1978, including prior approval on retransfers and reprocessing (of material irradiated in U.S.-supplied reactors or of U.S.-supplied fuel), no "peaceful" explosives, timely warning criteria, renegotiation of existing agreements, and full-scope safeguards. The "timely warning" question turned out to have major implications for what was, until this point, conventional wisdom about technological change in the nuclear fuel cycle. The issue here was whether ordinary IAEA safeguards could be politically relevant when dealing with plutonium fuels or sensitive facilities. It was argued that a primary role of safeguards is to provide a signal to the world that diversion or misuse is occurring in time for international response to be mobilized prior to consummation of nuclear-weapons construction or an irreversible commitment to weapons. The timely warning condition was intended to ensure that this function of safeguards remained intact and put the burden of proof on those proposing new fuel-cycle activities. This viewpoint was later to form a central pillar of the Carter nonproliferation policy.

The only significant piece of legislation *passed* by the 94th Congress was the Symington Amendment to the Foreign Aid bill. The amendment provided for a cutoff of U.S. aid to those countries importing or exporting enrichment or reprocessing equipment without guarantees of full-scope safeguards. This condition was made subject to presidential exception, with congressional override possible—again a model for the procedures of the Nonproliferation Act of 1978.

Foreign supplier exports were also the subject of a proposal, made prominent by Senator Abraham Ribicoff,[18] to implement reactor market sharing as a way to avoid the competition that led to inclusion of sensitive technologies in reactor sales in an export market increasingly centered in developing countries. Ribicoff also proposed the internationalization of spent-fuel storage and, to undercut arguments for reprocessing, the provision of low-enriched uranium fuel assurances. The Ribicoff proposal thus continued the trend toward increasing emphasis on technological choice as a determinant of proliferation risks; it identified commercial competition as part of the dynamics of proliferation; and it proposed low-enriched uranium fuel assurance, in a multilateral context, as a way to alleviate the pressure for commitments to plutonium. Anticipating future U.S. policy shifts, Ribicoff also suggested that the fuel assurances that were to be used as a carrot in the developing world could also be used in industrialized countries as a stick to compel conformity of foreign suppliers to U.S. nonproliferation goals.

Since very little of the legislative and other activity of 1976 resulted in actual changes in U.S. policy, there was little direct cause for alarm abroad. However, lack of clarity in U.S. export procedures, growing pluralism and dissonance in the policy formulation process, and indications of incipient changes in the basic assumptions and modes of action underlying U.S.

policy, tended to increase uncertainties, in major industrialized countries and developing countries alike, about the market and sovereign costs that would be associated with future supplies.

These issues came to a head in late 1976 and early 1977. The growing debate over the efficacy of U.S. nonproliferation policy found expression in the presidential campaign, and changes in the 95th Congress cleared the way for legislative action. In the later days of the campaign, the Ford administration promulgated a new policy regarding reprocessing and recycle of plutonium in which such technological extension of the fuel cycle would be considered necessary only when economic or other benefits outweighed proliferation risks. While this was a relatively conservative statement compared to those emerging in congressional debates or originating in the arms control community, it established an unusually strong linkage between domestic nuclear policies and foreign policy objectives.

It also marked a fundamental change in the basic assumption that there was no strong connection between technological change in nuclear power and the nature of the proliferation problem. While the United States had avoided direct transfers of enrichment and reprocessing technologies, the international nonproliferation regime it had helped to construct assumed that safeguards would be able to deal with technological change. The policies of France and Germany—which led to the deals with Brazil, South Korea, South Africa, and Pakistan—were consistent with the old assumptions but not with those emerging (in part as a result of these transactions) in the United States.

The Carter administration carried the debate further in a major announcement of April 7, 1977.[19] Domestic reprocessing and recycle of plutonium were deferred indefinitely and the commercialization phase (though not longer term research and development) of the breeder reactor program was suspended. Alternative fuel cycles, which inhibited access to weapons material, were to be emphasized in U.S. programs; fuel assurance was to be improved by increasing U.S. enrichment capacity and making enrichment contracts available; the export embargo on enrichment and reprocessing technology would be continued; and the United States would explore ways to ensure adequate energy supplies multilaterally while reducing the spread of capabilities for nuclear explosive development. The president also called for an international nuclear fuel-cycle evaluation program.

The focus of these statements, including that on fuel assurance, was on the proliferation implications of reprocessing and plutonium utilization. While the announcement indicated that the United States would not attempt to force abandonment of reprocessing in countries already having reprocessing plants in operation (Japan, France, Great Britain, and Germany were mentioned), the U.S. attitude toward the much larger number of countries with reprocessing *plans* (some predicated on waste-management assumptions) was far less clear. In addition, there was a suggestion (in response to a press question) that supply of fuel by the United States could be used as an instrument of compulsion as well as assurance.[a] In subsequent actions, the United States initiated the renegotiation of Agreements for Cooperation to reflect the new conditions.

Abroad, responses to the new policy were vigorous and largely in opposition. There were several reasons for this. Until the Carter statement, U.S. proposals for more restrictive nonproliferation conditions—which had included virtually all of those finally endorsed, and some more extreme— had been raising suspicions and concerns outside the United States. But they had not been carried to implementation and thus confrontation. Even the issues raised during the presidential campaign could be viewed casually, and with some skepticism, as campaign rhetoric. The final statement of a firm position thus catalyzed and focused reactions to trends that had been building for some time. In part because of a lack of precision in the announcement, the interpretation of the U.S. statement was frequently more extreme than was warranted; for example, the trade press and foreign nuclear officials often saw the United States as attempting to deny reprocessing and breeders everywhere. Such a denial would threaten long-standing plans for enhancing long-term energy security in some of these countries.

The early response came primarily, and most vehemently, from those directing nuclear programs in foreign countries or formulating national energy policies (often the same individuals). The new U.S. policy questioned the basic assumptions (the energy efficiency and independence provided by plutonium, a putative need for reprocessing for waste management, and so forth) on which programs and policies were built, and this happened at a time when domestic public opposition was beginning to be felt in those countries. Some nuclear-supplier states were put in a difficult position by the new U.S. policy. Germany, on the one hand, did not want to retreat on its contractual obligations to Brazil; on the other hand, proceeding in the face of U.S. opposition risked other foreign policy costs and a strengthening of domestic nuclear opposition.

Nuclear planners in some smaller countries also saw the shift as reshaping the future nuclear fuel market, forcing reliance on a few suppliers of enrichment and reprocessing services rather than allowing competition to emerge (for example, there were reports of Swiss concern about depending on France for reprocessing[20]). More importantly, others, such as Spain and Yugoslavia, saw the demand for renegotiation of existing agreements as a new sign of unreliability on the part of a traditional supplier of fuel and

[a]The statement is somewhat unclear: "If we felt that the provision of atomic fuel was being delivered to a nation that did not share with us our commitment to nonproliferation, we would not supply that fuel."[19]

technology, and as a further threat to the sovereignty of smaller, less-developed countries already highly dependent on large industrialized states. This issue was already one of concern to developing countries in the context of the NPT, and more generally in the debate over the relationships of rich and poor nations. By emphasizing the need to slow the spread of sensitive nuclear technologies beyond a few industrialized countries, the United States was adding another level of discrimination among countries—that is, all would not have equal freedom of technological choice in nuclear power development. Non-weapons countries had seen the NPT as offering free access to technology and fuel, with safeguards being the price to be paid. Now, the United States was proposing to limit technological access, but offering a countervailing benefit in the form of improved fuel assurance.

The indefinite deferral of domestic use of plutonium in the United States, based on a lack of economic or other need, was expected to minimize the appearance of discrimination. The U.S. shift was thus interpretable abroad as an effort to redefine the NPT bargain, an effort not obviously contrary to the interests of many developing countries but politically difficult to accept.

During this same period the new Congress was completing a committee reorganization that abolished the Joint Committee on Atomic Energy and established the exclusive oversight of international nuclear policy in the House International Relations and Senate Foreign Relations Committees. This step, and the emergence of an administration position, led eventually to the passage of the Nuclear Nonproliferation Act of March 1978 (NNPA).[21] The terms of this act were the result of a process of compromise between congressional attitudes and those of the departments and agencies of the executive branch. The major difficulty was to find a way to establish uniform conditions for nuclear exports, which would make approval predictable as soon as basic nonproliferation conditions were met, while preserving flexibility for the executive in dealing with specific situations. To provide this, the act specified sequential and conditional procedures involving the various governmental actors: the NRC, the president, the Departments of State and Energy, the Arms Controls and Disarmament Agency (ACDA), and committees of Congress.

In order to qualify for American-supplied fuel, an importer must have an Agreement for Cooperation, negotiated by the Secretary of State with participation by other governmental entities. In addition, specific shipments of nuclear fuel require an export license provided by the NRC. The 1978 act specifies new requirements to be included in Agreements for Cooperation (existing agreements must be renegotiated) and designates similar criteria to be applied by the NRC when considering an export license application.

The requirements for Agreements for Cooperation are set out in an amended Section 123 of the 1954 Atomic Energy Act. They include: 1) safeguards on all exports and material produced with exports; 2) full-scope IAEA safeguards—safeguards on all the peaceful nuclear activities of the recipient country; 3) a pledge not to use any material or equipment exported for research into or detonation of an explosive device, or for any other military purpose; 4) the U.S. right to require return of any exported material in the event of an explosion or abrogation of an IAEA safeguards agreement; 5) U.S. prior consent before retransfer, reprocessing, enrichment, or storage of any exported material; 6) adequate physical security measures; and 7) a guarantee that any facility built using technology transferred under the agreement would be subject to similar conditions.

Many of these conditions can be exempted by the president for foreign policy or security reasons, though the president is directed to renegotiate all existing Agreements for Cooperation to incorporate the new antiproliferation restrictions. Various committees of the Congress then have the opportunity to review the agreements.

The 1978 act also adds two other amendments to the 1954 Atomic Energy Act, containing criteria governing U.S. nuclear exports. These criteria—which are to be applied to each specific export license by the NRC—are virtually the same as those to be included in the Agreements for Cooperation, except that only the full-scope safeguards requirement (number 2 above) can be waived by the president, under special conditions and (until a 1983 Supreme Court decision declared the procedure unconstitutional) subject to congressional disapproval. Thus, while the provisions in the Agreements for Cooperation are subject to negotiation between the United States and the recipient party, the criteria governing license application are almost all mandatorily imposed by the United States, the only exception being the full-scope safeguards requirement. The process of obtaining an export license exemption on the full-scopes condition, starting with a presidential decision, is a laborious one, involving several congressional committees and further consultations with the Departments of State and Energy, ACDA, and the NRC. The entire export procedure has been clarified by the act, but the strong antiproliferation policy, and the provisions for involvement by disparate parts of the U.S. government, do not necessarily serve to relieve uncertainty about the outcome in particular instances.

In regard to fuel supply, a basic question about the act, raised by a number of countries, was whether it provided a clear and predictable export policy: many of the situations to which it applied did not satisfy the general conditions, and recourse to the exemption procedures would be required. This procedure introduced less predictable factors, involving action, and possible reactions, by the NRC, the president, and the Congress. It was thus difficult for some countries to regard the act as providing much greater assurance of supply, especially those which have not signed the NPT or accepted full-scope safeguards. Ironically, some of these were countries in which assurance may be most important as a nonproliferation measure.[22]

Indeed, by creating greater uncertainty—or lack of assurance—for countries that do not accept all of the basic nonproliferation conditions, the NNPA is perhaps more easily seen abroad as a tool of the U.S. drive for more stringent nonproliferation conditions, than as streamlining and making export conditions more predictable. Perhaps attempting to compensate for lack of manifest improvements in assurance through export policies, the act attempted to create a separate institutional mechanism to deal with fuel-assurance concerns, an International Nuclear Fuel Authority. The extent to which such a mechanism could deal with fuel-assurance concerns—especially those of countries whose primary uncertainty was U.S. export policy—was not clear and it has not been implemented.

Multilateral and Other Responses. Changes in the terms of nuclear trade, and thus the environment affecting nuclear fuel-supply conditions, were not limited to the United States, Canada, and Australia. The export policies of Germany and France, the predominant new suppliers of nuclear technology, were altered subsequent to their entry into export markets in 1975 and 1976. In general, these policies became more restrictive than at the outset, especially in regard to transfer of proliferation-sensitive technologies. This change was the result of internal policy changes stimulated by international discussions and pressures. In Germany, a traditional business and trade orientation in export policy was increasingly tempered by greater sensitivity to domestic and international political factors. In France, the creation of a Council on Nuclear Foreign Policy in 1976 similarly integrated political factors into an export policy that had, until then, been dominated by the Ministry of Trade and Industry.

In addition, as noted in Chapter 1, the London Suppliers' Group has agreed to increasingly restrictive common export conditions on nuclear trade and has provided an opportunity to discuss nonproliferation issues. According to press and other reports, the Suppliers' Group has also considered proposals by the Soviet Union and the United Kingdom for more restrictive supply conditions, including a suggestion that items on the IAEA "trigger list" of sensitive materials and equipment be provided only to countries accepting full-scope IAEA safeguards on all fuel-cycle activities.[23,24] France has opposed such a condition on the grounds that it would be equivalent to signing the NPT. While French policy shifted toward greater prudence in exports and willingness to discuss export issues, it appears (as in the 1976 communique of the Council on Nuclear Foreign Policy) to favor emphasis on bilateral rather than multilateral arrangements—a position that put a high value on independence and sovereignty for supplier and consumer alike.

This problem of sovereignty, and the difficulty of persuading reluctant allies and developing countries to alter their nuclear development plans, was recognized by the Carter administration in its April 1977 call for an international re-examination of nuclear fuel-cycle issues. The result was an organizational meeting of representatives of forty nations and four international organizations in Washington in October of 1977. This conference launched the International Nuclear Fuel Cycle Evaluation (INFCE), a technical and analytical study in eight major areas, including fuel-supply conditions, reprocessing, fast breeders, waste management, and advanced fuel cycles. The INFCE was explicitly not to be a negotiation and its stated objective was to pursue nuclear power development in a manner consistent with nonproliferation. It was also to pay special attention to the needs of developing countries.

While much of the INFCE activity was technical in nature, the effort accomplished several important nontechnical objectives. It effectively defused the intense international battle over nonproliferation policy in a way that respected the sovereignties of the nations involved, while encouraging nations to examine their own policies and technology decisions in an explicitly international environment. In doing so, it bought time and created opportunities for quieter efforts of diplomacy. As it turned out, the delay also suspended critical decisions until new trends in international nuclear fuel and technology markets had become clearer.

MARKET ADJUSTMENT AND
POLICY ACCOMMODATION

By the time the INFCE activity came to a close in February 1980, it was becoming evident that the market stresses of the earlier years were lessening. Uranium production capacity worldwide had doubled from its 1976 level and prices had just begun to decline; enrichment-plant expansion in Europe was close to eliminating the previous near monopoly of the United States; and nuclear growth plans were being revised downward in response to reductions in expected electricity-demand growth rates and social and regulatory restraints. These changes, explored in subsequent chapters, reduced the level of sovereign risk perceived by consumer nations, relieved the pressure for immediate decisions on commitments to new nuclear technologies, and encouraged supplier nations to reconsider their export policies in the light of new commercial imperatives. More generally, the confrontation over nonproliferation principles gradually evolved into an era in which other policy objectives weighed more heavily.

The first major changes came in Australia, which had missed benefitting from the uranium boom of the 1970s but which had rich new uranium ventures working their way toward production. Where earlier governments had effectively inhibited exports while domestic and international issues were being examined, the new Fraser government acted in May of 1977 to begin the implementation of an export-oriented uranium policy. It was encouraged in this by the results of the Fox Commission enquiry (see Chapter 5) and by the U.S. emphasis on low-enriched nuclear fuel assurance as a way to relieve the pressures for early commitments to plutonium.

While new bilateral agreements were negotiated with potential customer nations during the conduct of INFCE, the crucial issue of how Australia might exercise a right of prior consent to reprocessing of material of Australian origin was left open, pending the completion of INFCE discussions. In November 1980, well after the close of INFCE activities and after the election of the Reagan administration, Australia presented a new approach to the prior-consent problem. In effect, Australia offered to give generic approval to reprocessing of spent fuel within the nuclear energy programs of her customer nations. This programmatic approach was in significant contrast to the U.S. NNPA approach which requires case-by-case approvals even for close allies.

The Australian agreements also avoided several of the problems experienced by Canada and the United States. Under the agreements, retransfers of material of Australian origin between nations having bilateral agreements with Australia are permitted (thus avoiding disputes over retransfer between Euratom nations) and transfer to nations not having such bilaterals is permitted as long as equivalent amounts of material are returned. The latter clause, for example, allows Finland to have Australian material enriched in the Soviet Union.

The new Australian policy thus embodies an element of mutual trust, or shared responsibility, that is extended to those non-weapons states accepting full-scope safeguards. In a sense, this policy returns to the assumptions of the pre-Carter era, in which the form of discrimination is not over technological choice but over the willingness to accept the NPT bargain.

The new Australian approach put pressure on Canada and on the United States to accept similar terms. For Canada, the end of the INFCE exercise re-opened the question of how to deal with Euratom over the prior approval issue. Under the agreement signed in December of 1977, the Euratom nations were to engage in prior consultation with Canada regarding retransfers and processing of Canadian material, but Canada—during this period—would not demand the right of prior approval. This agreement called for negotiations to begin late in 1979 but this date was allowed to slip. When the agreement finally expired at the end of 1980 without negotiations having occurred, it was extended for another year, well beyond the end of the INFCE activity.

In the interim, in March 1981, Australia reached agreement with the European Community on generic, or programmatic, approval of reprocessing and retransfers. Following this, in November, Canada and Euratom initialed a final agreement that represented a significant but quiet retreat from the conditions originally demanded, with what was essentially a renewal of the conditions of the 1959 agreement between Canada and Euratom. Like Australia, Canada gave generic approval to reprocessing within the European Economic Community (EEC), with only a requirement of notification if there were major changes in nuclear programs.

The settlement of disputes over Australian and Canadian conditions on supply left only the United States as the primary source of contention over nuclear fuel trade. Under the NNPA of 1978, the president has the right to exempt exports to particular nations from the prior-consent clause affecting reprocessing, for a renewable period of one year, if failure to grant such an exemption "would be seriously prejudicial to the achievement of United States non-proliferation objectives or otherwise jeopardize the common defense and security."[21]

Because of difficulty in negotiating new agreements with Euratom nations, President Carter authorized such an exemption in 1980 and this exemption was renewed in 1981, 1982, and 1983 by the Reagan administration. In effect, these waivers mean that Europe is largely free of the constraints imposed by the NNPA and thus, as long as renewals are forthcoming, these nations are under little pressure to renegotiate their agreements with the United States to incorporate the stronger conditions of the act. The Euratom agreement itself expires in 1995.

A policy of continuous exemptions, while anomalous in the light of congressional intent, is perhaps more congenial to the outlook of the Reagan administration—and, certainly, to U.S. allies—than it was under Carter. In a major foreign policy statement on nonproliferation on July 16, 1981, President Reagan outlined a basic—if less than specific—position for his administration.[25] In this statement, support of basic nonproliferation goals was stressed, including a desire to prevent the spread of nuclear weapons, to view any nuclear explosion by a non-weapons state as a threat to international security, and to inhibit the transfer of sensitive nuclear material and technology where there is danger of proliferation. Support was also given for the NPT, the Treaty of Tlatelolco (which creates a nuclear-free zone in Latin America), and for IAEA safeguards.

All of these goals represented a continuity with previous U.S. policies. But what changed were the means to be used in achieving these goals and relative emphases on nonproliferation and other foreign policy objectives. Emphasis was placed on improving regional and global stability and reducing motivations to acquire nuclear weapons, on international cooperation, and on use of nuclear trade and nuclear cooperation as a tool of foreign policy. According to the background fact sheet accompanying the president's address: "This shift of emphasis from the previous Administration means that increased recognition will be given to the fact that proliferation is an international political and security problem, and not just a matter of control on the civil nuclear fuel cycle."[25]

Reinforcing this view was the president's statement that his administration would "not inhibit or setback civil reprocessing and the breeder reactor development abroad in nations with advanced nuclear power programs where it does not consitute a proliferation risk." He also remarked that "many friends and allies of the United States have a strong interest in nuclear power and have, during recent years, lost confidence in the ability of our nation to recognize their needs."[25]

In a sense, this shift represented a swing back in the direction of the philosophy of the original "Atoms for Peace" program and away from the Carter administration view (and that of the NNPA) that some technologies were simply too dangerous to be pursued, unless there were major advances in safeguards that would give "timely warning" of diversion. However, in returning to the old vision of the peaceful atom, the new administration finds itself forced to confront an old problem of discrimination (as well as the new limitations of the 1978 act). The Carter administration had been unwilling to make distinctions between nations, the logical consequence of which was that if it did not want plutonium to be used in developing countries, the world would have to forego its use elsewhere. While the Carter administration justified this change of technological strategy on economic and other grounds, not all industrialized nations were able to accept this reasoning.

In contrast, the Reagan administration's desire not to twist the arms of U.S. allies, who do "not constitute a proliferation risk," raises the question of how it will treat nations where, by the implicit distinction, there may be such risk. Here there are a number of problems. A general policy of selective approvals will leave the industrialized nations more vulnerable to claims of discrimination, in conflict with NPT obligations as perceived by the developing world. In addition, there is a question about whether it will be possible to encourage pursuit of plutonium fuel cycles and transfers of enrichment capacity to other industrialized nations, without significantly increasing the chances that the desire for such technologies, and the likelihood of their transfer, will increase in developing or potentially unstable regions.

This concern may explain why the Reagan administration has apparently encouraged new meetings of the Zangger Committee (but not of the Suppliers' Club) to extend safeguards associated with technology exports. Meetings late in 1982 and early in 1983 were aimed at expanding the "trigger list" to include centrifuge enrichment parts and equipment. Future meetings are likely to include items relating to reprocessing and other enrichment technologies.

But better indications of how the Reagan administration weighs non-proliferation in its foreign policy agenda come from how it has handled particular situations. These include problems associated with U.S. supply agreements with India and South Africa, relations with Argentina and Pakistan, and the reaction to the Israeli bombing of Iraq's nuclear research facility.

In the case of India and South Africa, the new administration inherited problems arising under the Nuclear Non-Proliferaction Act that had also proven difficult for the Carter administration. Fuel for the Tarapur reactors had been an issue since the passage of the 1978 act, which India claims was a unilateral abrogration of its thirty-year agreement for cooperation with the U.S., signed in 1963. This agreement specified that only U.S. material be used in the Tarapur reactors, while the act demanded that India accept full-scope safeguards before receiving new supplies, a demand that India refused. While the Carter administration had narrowly obtained congressional support for fuel shipments to India under waiver provisions of the 1978 act, congressional opposition subsequently intensified. In an agreement arranged by the Reagan administration and signed November 1982, France and India agreed that France would act as substitute supplier for the United States under the terms of the original 1963 agreement, rather than under the new NNPA restrictions.

A similar arrangement was worked out for South Africa. Original agreements and enrichment contracts called for the United States to be the sole supplier of enriched uranium for South Africa's Koeberg reactors. However, because South Africa has not signed the NPT or accepted full-scope safeguards (though it did accept safeguards on commercial reactors), the United States, under the terms of the NNPA, could not meet these delivery obligations. As a substitute, South Africa in late 1981 purchased enrichment services in Europe with the assistance of a U.S. broker and resold its U.S. enrichment services to a Japanese utility. While it does not appear that there was official U.S. government involvement in this transaction, the executive branch did not object (though some members of Congress did so). It is evident from this and other transactions that in a surplus nuclear fuel market, the United States has lost most of its fuel-supply leverage, and commercial mechanisms that circumvent political constraints are readily available.

In other areas, the Reagan administration also appears consistently to weigh general foreign policy objectives against direct nonproliferation controls. For example, the administration sought and won congressional approval for a massive conventional military-aid program to Pakistan. While it might be argued that such a move would reduce pressures on Pakistan to develop nuclear weapons, Congress had in the past refused to authorize aid to nations thought to be developing nuclear-weapons capabilities. Such resolve was evident in the Symington Amendment of 1976, discussed earlier, which itself had been partially motivated by France's sale of a reprocessing plant to Pakistan. The Reagan action seemed to be motivated not so much by a different view of how to prevent proliferation as by a desire to counter the regional Soviet threat already expressed in Afghanistan.

The administration has also seemed willing to risk questioning the institutional basis of the international nonproliferation effort. In June 1981, Israel bombed and destroyed an Iraqi nuclear research facility, contending that Iraq was engaged in a nuclear-weapons program. While there was some justification for Israel's suspicions (including technology tranfers from France and Italy, and a sale of more than 300 tons of uranium from Portugal for which little legitimate use could be foreseen), Iraq was an NPT signatory while Israel was not. This precipitated a confrontation in the IAEA in September 1982 when a close vote resulted in rejection of Israel's IAEA credentials. The United States walked out of the meeting and subsequently announced that it was reassessing its role in the agency. While few doubted that the United States would ultimately return, as it did nearly five months later, support for Israel and objections to what was referred to as "politicization" were apparently deemed more important than consistent support of the principal multilateral nonproliferation institution.

More recently, the administration has sought to improve relations with Argentina, as it did earlier with South Africa, based on what are perceived to be larger geopolitical factors. In this atmosphere, approval was granted in August 1983 by the U.S. Department of Energy for retransfer from West Germany to Argentina of heavy water of U.S. origin. In addition, the State Department called for timely review by the Nuclear Regulatory Commission of license for direct exports of nuclear power components to Argentina.

This pattern of administration actions is consistent with the view expressed in the president's statement in 1981 that proliferation was just one dimension of national and international security concerns. Subsequent actions suggest that it will only occasionally be the dominant consideration. Of course, this was also true at times in the Carter years; however, the relative priority of nonproliferation then seemed much higher.

Of fundamental importance to future U.S. actions is the increasing divergence between the current administration view and those of some members of Congress, especially as these views are expressed in the NNPA of 1978 and, potentially, in new legislation. According to these views, nonproliferation is an issue that transcends most other foreign policy objectives, and it is one that can be pursued through unilateral U.S. actions such as nuclear export policy, or military aid and economic assistance programs.

Where administration spokesmen have spoken of granting generic or programmatic approvals for the activities of U.S. trading partners as a basis for U.S. nuclear-fuel use and transfer approvals, congressional nonproliferation leaders refer back to the case-by-case and timely warning tests of the NNPA and speak of "tightening up" on nuclear and related exports and approvals. Such a divergence between administration and congressional views sets the stage for another era of uncertainty and volatility in U.S. nuclear policy. In a sense, this political divergence creates conditions that are at some variance with those frequently perceived from outside the country as being associated with the Reagan nuclear policy. Thus while the administration is able to offer encouragement to foreign trading partners—just as it offers encouragement to domestic nuclear endeavors—it may be essentially unable to resolve the problems of either.

NOTES

1. U.S. Department of Energy. 1983. *Statistical Data of the Uranium Industry.* Grand Junction, Colorado: U.S. Department of Energy. GJO-100(83). Table A-2 and Figure A-1.
2. *Nucleonics Week.* May 13, 1971. "Canada, Australia Move Forward Uranium Price, Processing Collaboration."
3. *Wall Street Journal.* February 8, 1972.
4. U.S. House of Representatives. November 4, 1976. *Hearings Before the Committee on Interstate and Foreign Commerce.* No. 94-150.
5. See, for example: Taylor, June H., and Michael D. Yokell. 1979. *Yellowcake: The International Uranium Cartel.* New York: Pergamon Press, pp. 121-122.
6. Author's estimate.
7. U.S. Atomic Energy Commission, Division of Production and Materials Management. 1974. *Survey of United States Uranium Marketing Activity.* WASH-1196(74).
8. U.S. Energy Research and Development Administration, Division of Production and Materials Management. April 1975. *Survey of United States Uranium Marketing Activity.* ERDA-24.
9. For a discussion of the claim of "commercial impracticability," see: Joskow, P.L. 1977. "Commercial Impossibility, the Uranium Market and the Westinghouse Case." *Journal of Legal Studies.* Chicago: University of Chicago Law School, no. 1: 119-176.
10. U.S. Energy Research and Development Administration, Division of Nuclear Fuel Cycle and Production. April 1976. *Survey of United States Uranium Marketing Activity.* ERDA 76-46.
11. U.S. Energy Research and Development Administration, Division of Uranium Resources and Enrichment. May 1977. *Survey of United States Uranium Marketing Activity.* ERDA 77-46.
12. U.S. Department of Energy, Division of Uranium Resources and Enrichment.

May 1978. *Survey of United States Uranium Marketing Activity.* DOE/RA-0006.

13. *Nucleonics Week.* April 22, 1976. "Iran wants an Option on a German-made Fuel Reprocessing Plant," p. 6.

14. *The Times* (London). July 13, 1976.

15. *Nuclear Fuel.* December 12, 1977. "Commercial Realities," p. 8.

16. *Nucleonics Week.* April 17, 1975. "NRC Finds Europe's 'Apparent Misunderstanding' of Export Licensing Policy Baffling," pp. 5-6.

17. For a detailed review of events in the United States during this period, see: Brenner, Michael J. 1981. *Nuclear Power and Non-Proliferation.* Cambridge, England: Cambridge University Press.

18. Ribcoff, A.A. July 1976. "A Market-Sharing Approach to the World Nuclear Sales Problem." *Foreign Affairs* 54, no. 4: 763-787.

19. Presidential Documents—Jimmy Carter. May 2, 1977. Vol. 13, no. 8.

20. *Nucleonics Week.* April 7, 1977. "U.S. No-Reprocessing-Anywhere Policy Denounced at German Meeting," pp. 5-8.

21. Public Law 95-242. *Nuclear Non-Proliferation Act.* March 10, 1978. (22 USC 3201).

22. For a discussion of the efficacy of fuel-supply leverage to achieve nonproliferation goals, see: Neff, Thomas L., and Henry D. Jacoby. 1979. "Non-Proliferation Strategy in a Changing Nuclear Fuel Market." *Foreign Affairs* 57, no. 5: 1123-1143.

23. *Nucleonics Week.* December 2, 1976. "The Toughest Nuclear Export Rules Yet Were Proposed By The USSR And The U.K.," p. 10.

24. *Nucleonics Week.* February 19, 1976. "Britain's Plan for Obtaining Wider Application of IAEA Safeguards," p. 7.

25. "Statement by the President." July 16, 1981. Office of the Press Secretary, The White House, and accompanying fact sheet.

3 URANIUM RESOURCES AND SUPPLY

Over the past decades, a great deal of attention has been paid to the assessment of uranium resources. But despite this attention, most assessments have either been flawed in important ways or have led to mistaken conclusions. Those seeking to justify research and development or demonstration projects for new nuclear technologies have usually seen rapidly depleting resources, requiring introduction of technologies that would use uranium and its byproducts more "efficiently." Those with experience in historic geologic environments—in the United States, Canada, and elsewhere—have had difficulty in seeing the potential for uranium in new environments, at home or abroad. Others have sought analytically to deduce mineral supply curves from data collected by the U.S. government or by the IAEA, but have failed to allow for the limitations of this data and its categorization, have extrapolated beyond these limitations, or have interpreted such limitations in data as real information about nature's limits. Geologists and mineral economists have argued over whose approach was correct or, taking public policy in their hands, disagreed about what was prudent. Finally, there has been chronic confusion between our immediate knowledge of uranium reserves and the long-run supply situation.

We shall not attempt here to review this long-standing controversy, as it has been done adequately elsewhere.[1] More importantly, much of the argument has focused on the United States and its traditional sandstone formations in the Wyoming Basin, the Colorado Plateau, and the Gulf Coastal Plain. However, discoveries outside the United States over the last decade or so have revealed a great richness of geologic expression and a supply horizon that recedes faster than our commitments to use of uranium fuels advance. It is, however, useful to consider the difficulties involved in resource assessment—by looking at extreme views at either end of the spectrum—as a prelude to examining the data that are systematically collected internationally.

RESOURCE PERSPECTIVES

Let us distinguish between two very different views of the underlying resource situation. In the first, suppose that uranium occurred in many different geologic environments as the result of a number of different processes and with a wide range of deposit sizes and uranium concentrations. Given perfect knowledge of the occurrence of all uranium, one could estimate production costs and derive a supply curve—the amount of uranium that would be available at a given price. According to this view, one would expect increasingly large quantities of uranium to become available as prices increased and as one thus had incentives to mine lower grades of ore, or deposits that were more difficult to exploit. Indeed, one might not even have to have great price increases if technical improvements in exploitation reduced costs as ore grades declined—the case historically with other minerals.

Alternatively, one might believe that uranium is discretely deposited in such a way that there is not a continuum of occurrences, grades, and sizes of deposit, but rather a limited quantity of moderate-grade material in a limited set of geologic environments. Once the most accessible of these deposits were identified, the cost of searching for and exploiting new deposits would rise rapidly, and so would uranium prices. This view is consistent with most of the facts about known sandstone deposits, say, in New Mexico.

Since we do not have perfect knowledge of the uranium resource endowment, we cannot *a priori* distinguish between the two extreme views above. However, the way in which information about uranium developed tended to bias some analysts in the direction of the second view. There are several ways in which this occurred. First, success in discovering reserves adequate to meet near-term demand (over the decade or so in which uranium industry investments are normally repaid) tended to inhibit further investment in developing additional information. If this success were achieved in one geologic environment or one region, there was little incentive to look elsewhere. Until recently, for example, most effort in the United States was focused on sandstone deposits in a few proven basins, and in Canada (which was intensively explored for weapons purposes) on particular formations in the eastern provinces. At least in Canada, the result of adequate success in the east was the failure to discover even richer deposits in the west.

Second, commercial interest in uranium is relatively new, and market conditions have been favorable to extensive investigation only over a brief recent period. Before the last decade, much of the exploratory effort was directed toward meeting the weapons needs of a few countries. The search was limited to countries friendly to the weapons states—Canada, Australia, South Africa, and the Belgian Congo (Zaire) for the United States and the United Kingdom; and former colonies in central Africa (Gabon and Niger) for France. Little effort was made elsewhere, and even in the producer countries in question, exploration was retarded by saturation of weapons requirements and declining real prices. For example, the *real* price (corrected for inflation) offered by the U.S. in 1955 was not achieved again—in the commercial spot market—until late 1975 (and not until even later for the majority of contracted supply); in the interim, it dropped by nearly a factor of four. Declining expectations led to conservative exploration and development behavior, with an emphasis on exploiting known low-cost reserves or looking nearby for deposits like those already being exploited.

Third, information development has significant lead times. While the existence of a new deposit may become known shortly after initial exploratory work (though the companies involved usually have strong incentives to restrict the availability of such information), it may be some years before enough is known to make estimates of reserves. Companies will invest in developing such information only in response to market signals; thus, additions to known reserves generally lag new demand indications. Overall resource and reserve estimates tend to lag even farther behind since time is required to analyze primary data and integrate them into a comprehensive view, especially at a world level. As a result, published estimates of national or global reserves and resources may lag by a decade or more the occurrence of the forces that motivated the exploration and other work leading to their discovery.

Bias in the magnitude of such estimates may also be introduced by the particular interests of the entities that prepare the figures. Companies and producer governments may have an incentive to take a conservative view of their uranium reserves—especially when there are large, rich deposits—lest the prospect of larger quantities undermine prices. On the consumer side, there is a complementary conservatism: utilities and consumer nations tend to take a worst case view of strategic energy commodities since the harmful consequences of overestimating availability are perceived as being much greater than those of underestimating it.

Thus, many of the forces influencing perceptions and knowledge of uranium resources and reserves work in the direction of conservative estimates and toward a view of ultimately limited resources. Typical of such estimates was that of the group of geologists comprising the Uranium Subpanel of the U.S. National Academy's Committee on Nuclear and Alternative

Energy Systems (CONAES) of 1978,[2] which displays several of the geologic conservatisms noted above. This view also reflects the strategic conservatism noted above: that one should not count, for planning purposes, on resources that are not already well known.

An alternative view was expressed in 1976 by Landsberg in the Nuclear Energy Policy Study sponsored by the Ford Foundation.[3] Here, the process of information development and resource exploitation is seen more in economic terms, and makes allowances for the tendencies noted above. It implicitly assumes the first view presented above of the resource situation, arguing by analogy with experience with other mineral commodities. Experience over the past few years lends strong evidence in support of this view. Discoveries in Saskatchewan, Australia, and elsewhere indicate a widening uranium-resource horizon. Large and very rich new deposits belie the claim that the world is moving toward an era of scarcity and rising costs. Instead we may well be moving into an era of lower costs. And discovery of very large lower-grade deposits (such as that at Roxby Downs in South Australia) suggests that the long-run supply curve will support greatly increased production rates with only moderate increases in prices. The details of this picture will be examined in Part II.

In part, the difficulty of assessing uranium resources arises from the changing state of geologic knowledge about this element. While uranium is frequently considered a rare mineral, it is in fact quite prevalent. The seawater concentration of uranium is 3.3 parts per billion, comparable to the concentrations of copper, nickel, and iron, and is several orders of magnitude higher than the concentrations of gold and other valuable elements. The average crustal abundance of uranium is about 3 parts per million.[4] Over geologic times, uranium was deposited in a number of environments at much higher concentration levels through a number of chemical and physical processes. Early efforts in the 1940s and 1950s emphasized exploitation of relatively low-grade deposits in sandstones and quartz-pebble conglomerates where concentrations range up to 0.3 percent, or about a thousand times average crustal abundance. More recently, deposits with concentrations up to 50 percent or more—a hundred thousand times average crustal abundance—and other deposits with very large quantities of uranium at commercially attractive grades have been found. This emerging picture is considerably more complex than was recognized in the early years of the uranium industry.

While uranium in the United States has been subject to intense study for several decades, much less systematic attention has been focused on uranium resources and reserves abroad. One result of this is not only less knowledge about prospects abroad but also a tendency to extend U.S. perspectives to non-U.S. environments. In fact, there are very significant differences. In what follows we seek to characterize world resources and put in context those of the United States.

Uranium appears in a number of different geologic environments.[5] Historically, some of the first deposits to be exploited were quartz-pebble conglomerates found at Elliot Lake and Agnew Lake in Canada and in the reefs of the Witwatersrand gold fields of South Africa. Uranium is relatively low in concentration in these deposits—perhaps 0.01 to 0.15 percent yellowcake, with the lowest grade economically recoverable only when coproduced with gold. These deposits are characteristically large, ranging up to 150,000 metric tonnes of contained uranium.

About the same time these conglomerates were being developed, uranium production began from sandstone formations in the western United States. These deposits resulted from special ground-water deposition conditions resulting in localizations of relatively low-grade ore, up to about 0.3 percent uranium oxide. Most of these deposits are small, though some have been found with as much as 40,000 MTU of recoverable reserves. Sandstone deposits have also been found in Niger (Arlit), Australia (Beverley and Yeelirrie), and in formations that underlie parts of South Africa, Zambia, Botswana, and Zaire.

The third type of deposit developed in early years consisted of veins of pitchblend or other concentrated mineralization, such as those at Port Radium (Canada) and Shinkolobwe (Zaire, formerly the Belgian Congo). These deposits can be high in uranium concentration, with an average minable ore of 0.1 to 2.5 percent yellowcake, but seldom involve more than 20,000 MTU of recoverable uranium. The United States has only one mine (the Schwartzwalder) exploiting a vein deposit.

Subsequent to the wartime procurements, several new types of deposits were developed to commercial status. Disseminated deposits in igneous and metamorphic rocks are found in Namibia, Brazil, and the Bancroft area of Canada. These deposits have relatively low concentrations of uranium—between 0.05 and 0.15 percent—but may be quite large. For example, the Rössing deposit in Namibia contains about 100,000 MTU at a grade ranging from 0.03 to 0.05 percent. Economies of scale and inexpensive chemical processing make exploitation feasible.

All of the previous types of deposits generally involve low grades of ore or small total recoverable quantities of uranium, or—in the case of U.S. sandstones—both. However, beginning in 1969 a series of discoveries made almost simultaneously in northern Saskatchewan and in the Northern Territory of Australia revealed the richness of what were characterized as "unconformity-related" deposits. Known examples of these deposits range in size from 10,000 to 170,000 MTU at uranium concentrations ranging from 0.2 to 40 percent or more.

There have also been significant discoveries in recent years in new mineral environments that do not easily fit into the five groupings classified above. The most significant of these is the Olympic Dam deposit at Roxby Downs Station in South Australia. While uranium is present only at concentrations of about 0.07 percent, the ore also contains large amounts of copper and significant quantities of gold and other rare minerals. Preliminary estimates put uranium reserves at a million tonnes. The Roxby Downs deposit was only discovered in 1977 and additional discoveries seem likely.

The worldwide picture is in strong contrast to the view one would form from looking only at U.S. reserves, or for that matter, at traditional producing areas in eastern Canada and South Africa. In the United States, there is a strong impression of a mature resource whose exploitation is inexorably proceeding to lower grades and higher costs. The picture outside the United States shows a highly varied resource whose lowest cost and largest deposits are only now being found. The contrast is even more striking when one observes the size and grade of deposits in the United States.

According to Department of Energy statistics,[6] reserves in the United States available at a forward cost below \$30 per pound of U_3O_8 (\$46 per kilogram U) total about 360,000 MTU. These reserves occur in 650 separate properties. The average ore grade is 0.10 percent U_3O_8 and the average property contains only about 500 MTU. While the distribution of deposits is quite broad, the twenty largest deposits contain an average of 6,500 MTU each at an average concentration of 0.09 percent oxide. For total reserves available below a forward cost of \$50 per pound (\$77 per kilogram U), there are 755 properties, the 53 largest of which contain more than 50 percent of the uranium. The latter have an average concentration of 0.07 percent and contain 6,000 MTU each. Similar statements apply to probable and potential resources, except that the uncertainty level is higher.

It is useful to compare these U.S. reserves with those identified outside the United States. To do so, we draw on the above discussion and on the analysis of specific mining ventures identified in Part II of this book. Not including Roxby Downs, these deposits hold a total of about 1.3 million MTU at cutoff grades that would be attractive at prices in the vicinity of those experienced in recent years. Some of these non-U.S. deposits are currently being exploited; all have been developed to the point where reliable estimates of reserves and ore grades may be made. In Figure 3-1, we show a scatter plot of these deposits, identified by the country in which they were found. There are of course many other smaller but commercially attractive deposits in other nations worldwide.

If U.S. deposits were to be plotted in Figure 3-1, they would virtually all lie in the lower left corner, below the 20 thousand tonne mark and to the left of the 0.20 percent mark. In contrast, most non-U.S. ventures lie outside this region. Indeed, Figure 3-1 is not large enough to include two groups of non-U.S. deposits. The first includes Key Lake and Cluff Lake in northern Saskatchewan. Key Lake contains about 70,000 tonnes of uranium in ore that has an average oxide concentration of about 2.8 percent. To include these deposits, Figure 3-1 would have to be more than five times wider. The

Figure 3–1.　Non-U.S. Uranium Deposits.

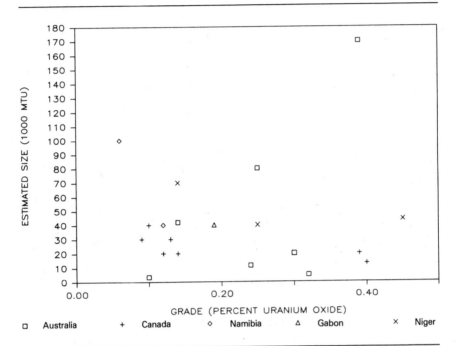

Note: Identified deposits for which reliable reserve estimates are available and where production is already occurring or might be initiated within a decade, given adequate incentives. Not included are Key Lake and Cluff Lake in Canada and Roxby Downs in South Australia, as well as a large number of smaller deposits worldwide. See Chapters 5 through 7, and Chapter 10, for details.

other type of deposit not shown is that at Roxby Downs in South Australia. If this relatively low-grade but very large deposit were included, Figure 3–1 would have to be more than five times as high as shown. While ore grades and deposit size do not entirely determine economic feasibility, the basic differences between a resource view limited to the United States and one taking into account worldwide conditions is clear.

TRENDS IN EXPLORATION
AND DISCOVERY

Given this background, we proceed to review trends in reserve and resource estimation, as conducted internationally. Outside the United States, the only consistent long-term series of estimates has been compiled by the OECD/

IAEA working parties on uranium resources.[7] Uranium estimates are based on country submissions, though working-party judgments are occasionally necessary. Two categories of resource certainty are considered:

- "Reasonably assured" for which there is direct quantitative geologic evidence of grade and quantity, say by drilling.
- "Estimated additional" for which there is direct geologic evidence of material, generally in association with known and delineated deposits.

These two categories are roughly parallel to the U.S. definitions of "reserves" and "probable potential resources." Until recently, the OECD/IAEA group did not estimate "possible" and "speculative" resources, as the United States has for many years. A rough comparison of reserve and resource categories as used by different groups is shown in Figure 3–2.

For each reserve or resource category, the OECD/IAEA group has separated estimates into high- and low-cost categories. These economic categories have undergone two changes over time—the first due to changing production costs and inflation, and the second due to a switch from a commercial price to a forward-cost basis (to be discussed below), in rough conformity to U.S. practice. Thus, early reports refer to uranium available at a *price* of less than $26/kilogram U ($10 per pound U_3O_8), while later reports refer to uranium available at a forward *cost* of less than $80 per kilogram U ($30 per pound U_3O_8). The switch to a forward-cost basis was intended to "maintain a certain stability in the resource categories employed and to become independent of possible significant price changes in the future which might be a reflection of the market situation and not necessarily be linked with actual recovery costs of a given type of resource."[8] Despite these changes, efforts have been made by the OECD/IAEA group to make historical estimates comparable. This effort appears to have been most effective for the reasonably assured (reserve) category. Changes in cost categories, and further efforts to convert potential resources to reserves, tend to move material into or out of the estimated additional category, either shifting into the reserve category, or into a higher cost bracket not considered in the estimates.

Table 3–1 shows the reasonably assured, or reserve, estimates made since 1967 (because escalation and inflation have moved material from one cost category to another, the two price categories are lumped together). Table 3–2 shows the corresponding figures for estimated additional resources. The largest shifts in and between categories are for Canada where the estimated additional category increased substantially in 1973 and the reasonably assured category declined in 1975. Apart from this anomaly, the estimates track reasonably well over time. Detailed data are tabulated in Appendix B.

Figure 3–2. Comparison of Terms Used in Major Resource Classification Systems.

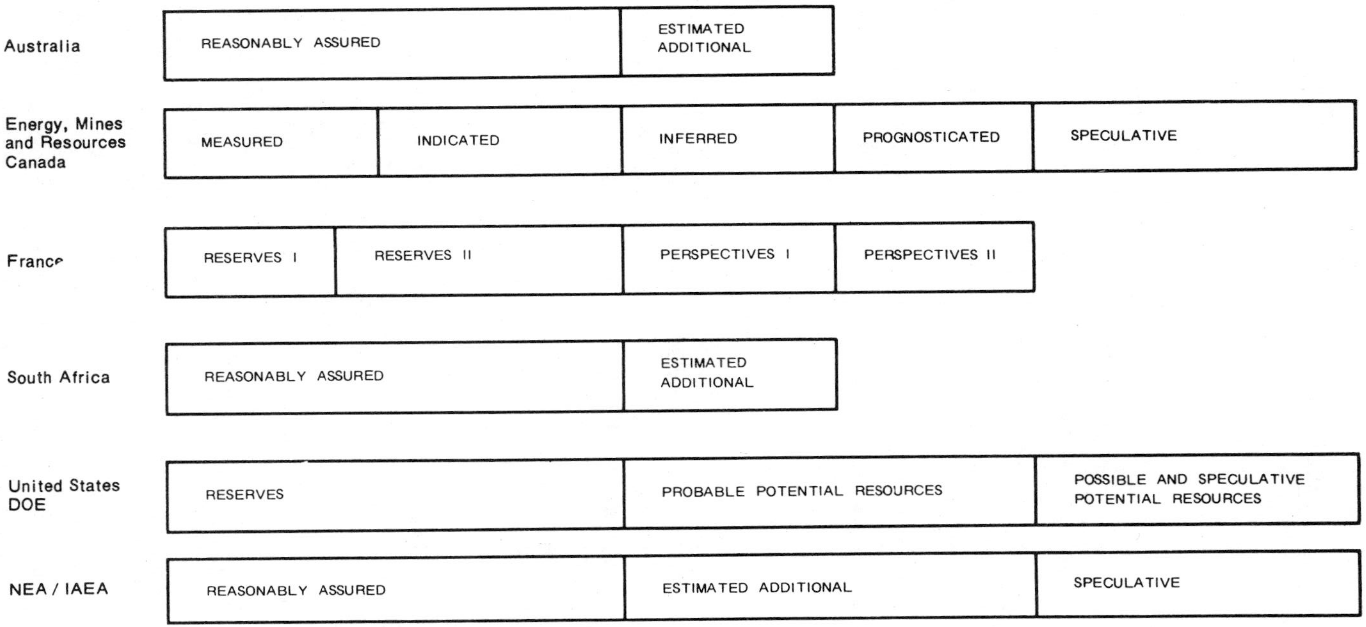

Australia	REASONABLY ASSURED	ESTIMATED ADDITIONAL	
Energy, Mines and Resources Canada	MEASURED	INDICATED	INFERRED	PROGNOSTICATED	SPECULATIVE
France	RESERVES I	RESERVES II	PERSPECTIVES I	PERSPECTIVES II
South Africa	REASONABLY ASSURED	ESTIMATED ADDITIONAL
United States DOE	RESERVES	PROBABLE POTENTIAL RESOURCES	POSSIBLE AND SPECULATIVE POTENTIAL RESOURCES
NEA / IAEA	REASONABLY ASSURED	ESTIMATED ADDITIONAL	SPECULATIVE

«The terms illustrated are not strictly comparable as the criteria used in the various systems are not identical. Grey zones in correlation are therefore unavoidable, particularly as the resources become less assured. Nonetheless, based on the principal criterion of *geological assurance of existence*, the chart presents a reasonable approximation of the comparability of terms».

Source: OECD Nuclear Energy Agency and the International Atomic Energy Agency. 1982. *Uranium Resources, Production and Demand*. Paris, France: Organization for Economic Cooperation and Development, p. 14.

Table 3-1.   Historical Estimates of Reasonably
Assured Reserves and Resources (*1,000 MTU*).

	Year of Estimate						
	1967	*1970*	*1973*	*1975*	*1977*	*1979*	*1981*
Australia	10	24	101	243	298	300	317
Canada	254	278	307	166	182	234	258
France	39	42	57	55	52	55	76
Namibia[a]	—	—	—	—	—	133	135
South Africa	208	204	264	276	348	391	356
Sweden	269	269	270	300	301	300	38
Other	68	105	150	322	370	451	503
United States	216	300	400	454	643	708	605
Total	1,064	1,222	1,549	1,816	2,194	2,572	2,288

a. Included in South Africa prior to 1979.

Source: OECD Nuclear Energy Agency and the International Atomic Energy Agency. *Uranium Resources,* 1967; *Uranium Production and Short-Term Demand,* 1969; *Uranium Resources, Production and Demand,* 1970, 1973, 1976, 1977, 1979, 1982. Paris, France: Organization for Economic Cooperation and Development. Detailed data are tabulated in Appendix B.

The reserve estimates are essentially those for uranium that would be available with some certainty on a relatively short time horizon, were there adequate demand. That is, these reserves are such that commercial mining investments and exploitation could be expanded within a few years (a period often set more by the need for environmental and other clearances than by construction times) following indications of sufficient demand. Indeed, we have been able to identify the location of deposits that total more than 80 percent of the quantities shown in Table 3-1. The reserves in Table 3-2 also represent the resource data that have been the longest time in preparation. The initiation of the exploratory and other efforts that resulted in additions to reserves late in the decade occurred early in the decade. The present reserves shown are thus the result of expectations about current and future markets as they were perceived five, ten, or more years ago.

Several important conclusions emerge from Tables 3-1 and 3-2. First, reserves and resources have increased significantly, in absolute terms and relative to prospective nuclear growth. Outside the United States (and excluding Sweden, whose shales are unlikely to be exploited in this century), reserves increased by a factor of 2.8 and estimated additional resources by a factor of 3 between 1967 and 1981. In absolute terms, non-U.S. reserves have increased by more than one million MTU; estimated additional resources have also increased by more than a million tonnes. These additions occurred during a period in which non-U.S. reactors required a cumulative total of only 85,000 MTU, and non-U.S. reactor requirements grew to

Table 3–2. Historical Estimates of Additional
Reserves and Resources (*1,000 MTU*).

	Year of Estimate						
	1967	*1970*	*1973*	*1975*	*1977*	*1979*	*1981*
Australia	3	10	108	80	49	54	285
Canada	354	308	409	419	656	727	760
France	23	31	49	40	44	46	46
Namibia[a]	—	—	—	—	—	53	53
South Africa	39	39	34	74	72	139	175
Sweden	38	—	40	—	0	3	44
Other	96	119	127	263	223	290	249
United States	404	621	769	812	1,053	1,158	1,097
Total	957	1,128	1,536	1,688	2,097	2,470	2,709

a. Included in South Africa prior to 1979.

Source: OECD Nuclear Energy Agency and the International Atomic Energy Agency. *Uranium Resources,* 1967; *Uranium Production and Short-Term Demand,* 1969; *Uranium Resources, Production and Demand,* 1970, 1973, 1976, 1977, 1979, 1982. Paris, France: Organization for Economic Cooperation and Development. Detailed data are tabulated in Appendix B.

an annual level of about 15,000 MTU.[9] In comparison with past and present reactor requirements, additions to reserves and known resources have been very large.

Perhaps more interesting is the comparison with estimated requirements forward in time. In 1967, known reserves and resources outside the United States stood at about 1.1 million MTU; at about this time, the estimate of non-U.S. nuclear power generating capacity ten or so years ahead (the 1969 estimate for capacity in 1980) was for 240 GWe.[10] This lead time is what realistically might be required to move substantial known uranium reserves and resources into production. According to this early estimate, growth in nuclear capacity would have required about 39,000 MTU[11] annually in 1980 (actual growth, of course, has been much less than this), for a ratio of reserves and resources to annual forward needs of about 30. That is, estimated reserves and resources in 1967 were about 30 times expected annual requirements eleven years ahead; alternatively, the resource time horizon was thirty to forty years off, if no additional reactors were considered.

According to the scenarios to be described in Chapter 4, non-U.S. uranium requirements for 1990 are estimated to be between 22,500 and 34,000 MTU annually. In 1981, reserves and resources were estimated at about 3.2 million MTU for a ratio of the latter to forward annual requirements of 90 to 140, several times those of a decade earlier.

Alternatively, estimated reserves and resources would provide more than a hundred years of forward supply for all reactors operating, under construction, or on order. Thus the uranium resource situation outside the United States relative to expected nuclear growth has improved significantly. Of course, the nuclear growth estimates play an important role in this: high early estimates may have stimulated exploration and development, while lower recent expectations mean that discovered resources will go farther. Not surprisingly, worldwide exploration investments have declined in recent years.

While uranium might have proven different from other minerals and energy sources, it is worth noting that known forward supplies of oil and of many minerals have—over much of this century—displayed resource horizons that indicated depletion at the then current consumption rates within, say, twenty or thirty years. Of course, the resource horizon has always retreated ahead of time and consumption; we have not run out. The reason is that knowledge about reserves and resources is usually gained only at some cost, and investments to produce this information will be made only if there is a prospect for a relatively near-term payoff. This term rarely extends beyond a decade or so, and for resource exploiters there is little need to know about material that might be needed beyond this horizon.

The cumulative record of investments in uranium exploration worldwide for the period 1973–83 is shown in Figure 3–3.[12] More than 55 percent of the uranium exploration expenditures made worldwide were made in the United States. According to Tables 3–1 and 3–2, the United States accounted in 1981 for only 26 percent of reserves and 40 percent of estimated additional resources. This comparison is, of course, overly simplified. One should allow for depletion, for real cost escalation (detailed data for exploration do not exist to do this), and other factors in any assessment of the effectiveness of exploration spending. For example, spending in the United States resulted in production of more than 277,000 MTU through 1980, in addition to the reserves shown in Table 3–1. However, spending in the United States also occurred before that in many other countries and, if expressed in terms of today's dollars, would account for an even greater percentage in Figure 3–3.

These observations suggest that greater care in assessing the data given here would not qualitatively change the conclusion that spending on exploration in the United States has become less productive than in some other nations, most notably Australia. As discussed earlier, the U.S. resource base appears to be at a much more advanced stage of exploitation than that even in other historic producer nations. In addition, it is evident from Figure 3–3 that exploration outside traditional uranium-producing nations has been relatively quite small.

Reserve estimates, detailed in Appendix B, also show a changing pattern of geographic origins for reserves and resources: major new additions have been made in areas that have not traditionally produced uranium. Between

Figure 3-3. Uranium Exploration Expenditures: 1973-1983 (*Totals in Constant 1972 U.S. Dollars*).

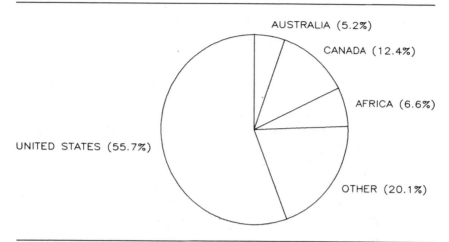

AUSTRALIA (5.2%)

CANADA (12.4%)

AFRICA (6.6%)

OTHER (20.1%)

UNITED STATES (55.7%)

Sources: OECD Nuclear Energy Agency and the International Atomic Energy Agency. 1977, 1979, 1982, 1983. *Uranium Resources, Production and Demand*. Paris, France: Organization for Economic Cooperation and Development; author's estimates where data are not reported. Expenditures in Australia for 1982, and estimate for 1983, are not available.

1967 and 1981, Niger's known reserves increased by about 140,000 MTU, an eightfold increase; Brazil's reserves—little evident before 1975—reached about 120,000 MTU in 1981; and Namibia's uranium reserves increased to an estimated 119,000 MTU. Overall, reserves in countries other than historic major producer nations (Canada, Australia, the United States, France, and South Africa) have increased by about 570,000 MTU between 1967 and 1981. New and prospective producers thus account for nearly 40 percent of the increase in reserves over the past decade. Moreover, countries previously not well explored have experienced major growth in reserves; for example, Australia's estimated reserves have increased thirtyfold since 1967. Similar statements apply to known resources.

These observations reinforce the suspicion that exploration and reserve definition—rather than geologic scarcity—are the principal barriers to knowledge of much greater uranium resources and reserves. Evidence for this also comes from major new discoveries in traditional producer countries such as Canada and Australia. Many of these discoveries are in new geologic environments and often have ore grades well above those already being mined. Some major discoveries, such as the reported million tonnes of uranium at Roxby Downs in Australia, do not yet appear in official estimates. This deposit alone is comparable to non-U.S. cumulative uranium consumption through the year 2000 for all presently committed reactors. The view that uranium is a mature resource whose exploitation is proceeding to lower and more costly grades and environments is thus contrary to the most recent evidence.

FORWARD COSTS AND
RESOURCE ECONOMICS

To date, there is no true economic theory of uranium as a worldwide resource. Such a theory would reflect exploration costs, an underlying resource model, and user costs (to account for depletion), as well as development investments (including appropriate discounting), transportation, milling and marketing expenses, taxes and royalties, and a host of other factors. Instead, what we have are rough indicators that are in fact much closer to engineering cost estimates for only the final stages of the exploitation process.

This lack of a developed theory appears to be partly the result of a history of government involvement in the earlier days of the uranium industry, partly due to the immaturity of the resource endeavor and the world market, and partly due to the recurrent dominance of political factors such as national defense, nonproliferation, and supply-security concerns. Initially, in the 1940s and 1950s the governments of weapons states and those of a few key producer countries promoted the development of uranium production and production capacity. Bonuses, guarantees, loans, infrastructural investments, and incentive prices motivated the discovery of uranium and the development of a major industry, especially in Canada, Australia, France, and South Africa. These involvements produced an economic environment and industry structure fundamentally different from what free markets would have produced. Later, as weapons demand fell, governments in producer nations acted to sustain domestic uranium industries. As it turned out, the nature of this later intervention had a major effect on how information about resources was gathered and analyzed.

The prices offered to producers during this period of surplus were universally lower than during the earlier period of liberal incentives. In Canada, government maintenance prices were only about half of those paid earlier; in the United States, a similar drop was recorded. In order to determine what prices should be paid and to set the quantities that would be purchased, the U.S. government sought information from producers about reserve holdings and about incremental production costs. This information eventually became the basis for U.S. resource evaluation, and the cost data gathered naturally led to the forward-cost concept of resource classification. Eventually, the forward-cost classification developed in the United States found its way—only slightly altered—into international resource assessments.

Forward costs include *future* variable and capital costs necessary to bring uranium to market. For a given deposit, such cost estimates may be used to decide what cutoff grade to use—that is, how much of the deposit to include in reserve or resource totals. In effect, forward cost is used both in the United States and abroad to decide which material is "economic." There are similarities and differences between the United States and international applications of the forward-cost approach: neither includes exploration nor investments made prior to the point of classification (which obviously comes at different points for different ore bodies); U.S. practice does not include return on (anticipated) capital investment though some account is taken of the cost of capital in international efforts; and inflation is accounted for simply by changing the cost categories or shuffling material between categories.

Several different forward-cost thresholds may be used, so that part of a deposit will be included in the lowest-cost category and the rest in another (of course, these forward costs are interdependent since the higher-cost material might not be considered if a mine were not dedicated to lower-cost material). Care must be taken in comparing U.S. estimates with those outside the United States: while foreign assessments are quite similar to those in the United States, a given forward-cost category in the United States is inclusive of lower forward-cost material, while those outside the United States are usually exclusive groupings.

Forward costs have frequently been interpreted as indicating the price at which uranium resources might be developed. Indeed, some have attempted to use data classifying uranium reserves and resources according to forward cost to develop a uranium supply curve. As discussed above, there are many problems with this approach. The first is that forward costs are just the incremental expenditures necessary to bring a given reserve to market, with no allowance for prior investments in exploration, mines, mills, or infrastructure, or for return on prior investments. A related problem is that forward costs may be very different for nearly identical reserves, depending on what investments have already been made. In addition, newly discovered reserves or resources about to be developed may have forward costs that are higher than those of existing but inferior deposits. Finally, necessary future capital investments are simply allocated over the total tonnage to be recovered and are not discounted to reflect the time value of money or adjusted for risk.

In using the forward-cost concept to measure reserves and resources, changes in mining costs, inflation, and other factors may lead to a "loss" of reserves or resources. Since forward-cost ceilings determine how much of a given ore body is assumed to be exploited (e.g., in setting the cutoff grade or assessing overall economic feasibility), the quantities of uranium apparently available will change even though there has been no real depletion or new physical information acquired. It is also important to remember that only a small fraction of the material in a given forward-cost category—say $50 per pound—would actually cost $50 to produce: there will be some material available at a forward cost of $35 and, in the United States at least (where higher-cost categories are inclusive of lower-), some at $12.

The forward-cost methodology is approximately consistent with a true economic assessment of resources and reserves only under very special conditions. One of these is the environment in which it was introduced: a period of surplus production capacity and weak demand in which market prices might be expected to approach variable costs of production over (what was expected to be) a short period of time before commercial demand picked up. The same might be true of the market conditions of the early 1980s. However, the forward-cost approach does not provide an adequate evaluative basis either for periods of rapid market growth in prices or quantities, or for longer-term resource development. In both of the latter cases, the attractiveness of new capital investment for long-term returns is critically important, and is not captured by the forward-cost approach. But even more importantly, the discovery and development of new resources is an ongoing process, while forward-cost categorization of reserves and resources represents but an imperfect snapshot of the status quo.

PRODUCTION HISTORY

Historically, uranium production worldwide rose to a peak in 1959 in response to weapons-procurement efforts. It subsequently fell with the decline of weapons demand and the delay of commercial nuclear power. During this period, producer governments instituted support programs in order to maintain a viable industrial base pending growth of commercial demand; in some cases, substantial stockpiles accumulated. These programs were most effective in the United States, especially in the initial years of commercial nuclear power growth, due in part to protectionist import restrictions exercised through the U.S. enrichment monopoly. It was not until 1979 that production reached the levels attained in 1959. The historic growth and decline of world uranium production are shown in Figure 3–4.

It is useful to compare historical production with actual reactor requirements during the period shown. Reactor requirements were computed from historical dates of initial operation for each existing reactor, assuming operation at 70 percent capacity factor and 0.20 percent tails assay. These assumptions probably overstate requirements: enrichment plants often ran at higher tails assay, but utilities delivered uranium as if tails assay were set at 0.20 percent (the additional uranium came from U.S. stockpiles rather than contemporary commercial sources), and capacity factors were well

Figure 3-4.   Historic Uranium Production.

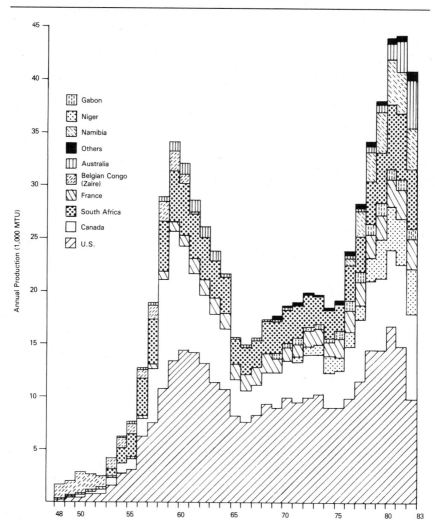

Source: Table B-1, Appendix B.

below 70 percent.[a] Annual reactor requirements outside the United States are compared with production in Figure 3-5. Since weapons requirements were dropping rapidly in the post-1965 era, much of the excess of production over commercial requirements was stockpiled, either by producers or

[a]A potentially countervailing factor might have been the failure of early fuel to achieve design burnup; however, this problem primarily affected early fuel loadings whose aggregate volume, compared to more recent consumption, was small.

Figure 3-5.   Non-U.S. Uranium Production Compared with U.S. AEC Foreign Purchases and Commercial Reactor Requirements.

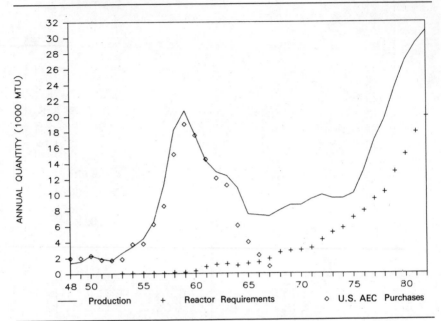

Sources: Production from Table 1, Appendix B; U.S. AEC purchases from Figure 2-1; and commercial reactor requirements computed from recorded nuclear power plant capacity additions using the fuel-cycle parameters of Table 1-1. Actual commercial purchases were significantly greater than reactor requirements with the excess being added to inventories.

by consumers. A similar history is shown in Figure 3-6 for the United States. In the United States and other weapons states, distinctions between military and non-military stocks were somewhat artificial. It is evident from Figure 3-6 that commercial stock accumulations in the United States may have been on the order of 100,000 MTU. Abroad, a similar quantity seems possible, held primarily by Canada, South Africa, the United Kingdom, and Australia.

The global supply and demand balance has actually been closer than suggested in Figure 3-5 because of enrichment contracting requirements, excessively high estimates of demand, and demand for inventories. The introduction of Fixed-Commitment contracts and new enrichment-plant feed requirements in 1973 tended—when compared with slipping reactor schedules—to result in uranium demand above that required for actual reactor use, as discussed in the next chapter.

As of January 1, 1983, U.S. utilities, reactor vendors, and other companies reported inventories of 59,100 MTU (natural uranium equivalent—

Figure 3-6. U.S. Uranium Production and Commercial Requirements.

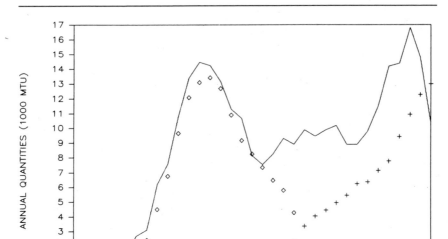

Sources: Production from Table B-1, Appendix B. AEC purchases 1956-71 from: U.S. Department of Energy. 1983. Statistical Data of the Uranium Industry. Grand Junction, Colorado: U.S. Department of Energy. GJO-100(83). Commercial requirements consist of actual reactor feed requirements, computed using the fuel-cycle parameters of Table 1-1 and recorded U.S. nuclear power capacity additions. Actual purchases by utilities were generally greater, with excess supply being added to inventories.

about 25 percent was already enriched). Non-military government-held stocks were reported, as of January 1983, being about 15,770 MTU of natural uranium and the equivalent of 44,270 metric tonnes of natural uranium held as enriched stock. Of total commercial inventories, 89 percent was held by utilities.[13] The pattern abroad varies and will be discussed in Chapter 8.

But the most striking feature of Figure 3-5 is its revelation of the relative immaturity of the commercial uranium market, compared to the length and magnitude of previous industry efforts. Reactor requirements were still less than half of production as late as 1974, and at the end of 1982 were still only about two-thirds of production levels. More material was purchased, but it went into already bloated consumer inventories. It is only in the past few years that requirements and supply have begun to appear to be related. One can thus argue that a mature commercial uranium market has only recently begun to emerge internationally.

REFERENCES

1. Adelman, M.A.; J.C. Houghton; G.M. Kaufman; and M.B. Zimmerman. 1983. *Energy Resources in an Uncertain Future: Coal, Gas, Oil, and Uranium Supply Forecasting.* Cambridge, Mass.: Ballinger Publishing Company.
2. Report of the Uranium Resource Group. Supply and Delivery Panel of the Committee on Nuclear and Alternative Energy Systems (CONAES). 1978. "Problems of U.S. Uranium Resources and Supply to the Year 2010." Washington, D.C.: National Academy of Sciences.
3. Keeny, Spurgeon M., Jr. 1977. *Nuclear Power Issues and Choices.* Cambridge, Mass.: Ballinger Publishing Company.
4. Driscoll, M.J., and F.R. Best, eds. December 1982. "Progress Toward the Recovery of Uranium from Seawater." Cambridge: Massachusetts Institute of Technology. Report no. MITNE-256.
5. For a detailed review of geologic conditions, see: OECD Nuclear Energy Agency, and the International Atomic Energy Agency. 1980. *World Uranium: Geology and Resource Potential.* International Uranium Resources Evaluation Project, Phase I Report. San Francisco: Miller Freeman Publications Inc.
6. U.S. Department of Energy. January 1983. *Statistical Data of the Uranium Industry.* Grand Junction, Colorado: U.S. Department of Energy. GJO-100(83).
7. OECD Nuclear Energy Agency, and the International Atomic Energy Agency. 1967. *Uranium Resources;* 1969. *Uranium Production and Short-Term Demand;* 1970, 1973, 1976, 1977, 1979, 1982. *Uranium Resources, Production and Demand.* Paris, France: Organization for Economic Cooperation and Development.
8. OECD Nuclear Energy Agency, and the International Atomic Energy Agency. 1982. *Uranium Resources, Production and Demand.* Paris, France: Organization for Economic Cooperation and Development.
9. See Figure 3–5.
10. See Chapter 4, Table 4–1.
11. Computed according to fuel-cycle requirements described in Chapter 1.
12. OECD Nuclear Energy Agency, and the International Atomic Energy Agency. 1982. *Uranium Resources, Production and Demand.* Paris, France: Organization for Economic Cooperation and Development. p. 44.
13. Energy Information Administration. Office of Coal, Nuclear, Electric, and Alternative Fuels. Nuclear and Alternative Fuels Division. September 1983. *1982 Survey of United States Uranium Marketing Activity.* DOE/EIA-0403.

4 URANIUM DEMAND

For many commodities, market demand is a relatively simple function of price and consumer preferences. If the price of the goods goes up, consumers buy less or switch to other, substitute goods. For many goods, such elasticities allow markets to equilibrate quickly between supply and demand, though there are often wide price swings. For uranium, the situation is even more complicated. The demand for uranium is the result of the complex fuel-procurement process described in Chapter 1, complicated by a host of market rigidities and by imperfections arising from political interventions. As a result, uranium has experienced larger price movements over relatively short periods than most commodities, other than oil.[1]

Demand for uranium at the reactor in the short run is almost completely inelastic. The utility operating a reactor has made a massive capital investment whose debt service accounts for a very high fraction of the cost of electricity generation. If the price of uranium goes up, the utility does not stop operating the plant, nor are there substitute fuels. Similarly, the fact that uranium is but a small part of the cost of nuclear power, and usually a small factor in decisions to build nuclear plants, means that even long-run reactor demand for uranium is relatively inelastic with respect to price. While there are minor technical elasticities possible—such as changes in in-core fuel management, longer burn cycles, or improvements in design—these factors are only of minor importance in assessing even long-term uranium market conditions.[2]

In subsequent discussions we shall regard demand *at the reactor* to be essentially inelastic, a simple function of nuclear-capacity growth that is not itself significantly dependent on uranium market conditions. Such a view is adequate for the next two or three decades. Beyond that, there are technical alternatives—plutonium and other fuel cycles—that *could* result in major shifts in the relationship between nuclear power growth and uranium demand. Such technological change might be accelerated by rapidly rising uranium prices or great insecurity of supply, but given our expectations about uranium market conditions, this appears quite unlikely.

These observations do not mean that demand for uranium as expressed in the market is closely related to reactor requirements. It is, in historical fact, quite different. In principle, there are several ways in which market demand may be different from reactor requirements. The first is through the way in which enrichment is used in the processing of nuclear fuel—a technical flexibility in which uranium and enrichment costs play a role. The second is that long-term and inflexible commitments to enrichment services—through long-term contracts or equity commitments—may dictate uranium purchases above reactor requirements. Finally, the desire for increases, or decreases, in inventories will alter market demand for uranium.

In the following sections, we first look critically at previous efforts to estimate nuclear power growth internationally. Based on this evaluation, we then select two scenarios that appear likely to bracket future growth. We then consider the effect on uranium demand of enrichment contracting under historic commitments to enrichment services by consumer nations. We briefly explore the role of inventories in uranium demand, reserving more extensive discussion to subsequent chapters. Finally, we consider the effects that technological change might have on demand over the remainder of this century.

ASSESSING NUCLEAR POWER GROWTH

Experience has taught us that it is very difficult to make accurate projections of future nuclear power growth. In Table 4-1, we present estimates made by the OECD and IAEA groups since 1969. These estimates include U.S. capacity and all other capacity outside the centrally planned economies (the latter we refer to as World Outside Communist Areas, or WOCA, throughout this book in keeping with industry conventions).[3] These estimates were compiled from reports by individual nations of their nuclear power plans and targets. Also shown are the actual capacities achieved. What is remarkable here is the continual downward revision of nuclear growth estimates and the apparent inability of forecasters to produce accurate estimates of future growth.

Table 4-1. Historic World Nuclear Growth Projections
(*WOCA Total in GWe*).

Year of Projection	Projection for Year:					
	1970	1975	1980	1985	1990	2000
1969	26	101–125	234–328	—	—	—
1970	18	118	300	610	—	—
1973	14[a]	94	264	567	1,068	—
1975		69[a]	194	530	1,004	2,480
1977			146	278–368	504–700	1,000–1,890
1978			144–159	243–272	374–460	834–1,207
1980			120[a]	232–258	361–401	585–804
1983				224	326	504–558

a. Actual capacity.

Sources: OECD Nuclear Energy Agency and the International Atomic Energy Agency. *Uranium Production and Short-Term Demand, 1969; Uranium Resources, Production and Demand*, 1970, 1973, 1976, 1977, 1979, 1982, 1983. Paris, France: Organization for Economic Cooperation and Development.

The reasons for the failure to make accurate projections of nuclear growth were several. To a certain extent, consistently high estimates were the result of the optimism of the nuclear proponents making such estimates. In more than one case, wishes or official targets were poor substitutes for realistic assessments. However, the last decade also saw basic shifts in both electricity demand and in the relative attractiveness of nuclear power. For many years electricity demand had grown exponentially at an annual growth rate of 7 percent in the United States, and often higher in other nations. With rising energy prices in the 1970s and recurrent recession worldwide, electricity demand has dropped to an annual growth rate of 2 to 3 percent.

While rising oil prices early in the last decade encouraged major new commitments to nuclear power in Japan, France, and other nations, as well as the United States, such commitments were subsequently revised downward. When the second oil price shock came in 1979, even this was not enough to retard the slide of optimism about nuclear power. With high oil prices came high interest rates, a critical factor for capital-intensive nuclear plants whose construction times were tending to lengthen due to slower than expected demand growth, safety questions and retrofits, and public opposition. Lengthened construction times meant higher financing costs and longer delays until returns on investments. In some electrical grids or service areas, electrical-demand growth rates declined to the point where the large capacity additions associated with nuclear plants were difficult to accommodate efficiently in either financial or physical terms. Given large uncertainties about future demand, it became reasonable to delay capacity decisions as long as possible and then, if necessary, build smaller fossil-fueled units.

Because of such difficulties in forecasting, we have chosen to use two different nuclear power growth scenarios to estimate future nuclear fuel demand. While rather mechanistic, these growth forecasts enable us to bracket a wide range of possible nuclear futures. The first of these scenarios is based on a survey of U.S. and non-U.S. utilities conducted by the American Nuclear Society's journal *Nuclear News.*[4] This survey asks utilities to report reactors in operation, under construction, or on order (where an actual order has been made or a letter of intent has been signed). Over the next decade, such estimates should give the maximum capacity to be installed (on average at least) since the time it takes to bring a new reactor into service from the date of order is now close to ten years in many countries. In fact, utilities have often been too optimistic about their ability to complete planned reactors, as indicated in Figure 4–1. Each year that a survey is made, utilities revise downward their estimates of *near-term*

Figure 4–1. Utility Estimates of Nuclear Capacity Expansion (*Non-U.S. WOCA*).

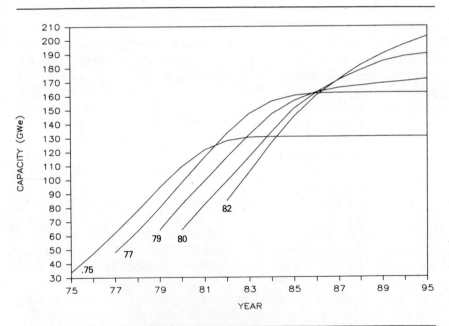

Source: Based on Nuclear News surveys of Non-U.S. utilities August 1975, August 1977, August 1979, February 1981, and February 1983. Only reactors in operation, under construction, or on order with a startup date specified are included. See Table B–3, Appendix B, for details.

capacity. But they also appear to increase the estimate of the number of reactors they expect to have operating more than seven or eight years in the future.

As one of our two growth scenarios, we have chosen to use the most recent (end 1982) utility survey[5] as the "Present Plans" growth scenario. A detailed listing, on a country-by-country basis, of this growth forecast is given in Appendix B. This growth scenario essentially assumes that no more reactors will be ordered beyond those already in operation, under construction, or on order (note however that some of these orders might be cancelled). This scenario thus gives us a base case of relatively firm demand against which to measure the effect of future reactor orders and the level of commitments to nuclear fuel supplies worldwide.

Of course, it is possible that more reactors will be ordered, at least in some nations. To take this into account, we have constructed a second nuclear power growth scenario. This "Moderate Growth" scenario assumes some near-term delay in plants now under construction or on order, but then envisions new reactor orders that will increase nuclear capacity and nuclear fuel demand above the Present Plans scenario beginning late in this decade. The Moderate Growth scenario is similar to projections made by the nuclear fuel-cycle services company (Nukem)[6] and by other groups. This scenario assumes that utilities outside the United States (non-U.S. WOCA) will order and install an additional 159 GWe of capacity before the year 2000 *in addition to* the 203 GWe now operating, under construction, or on order. We believe that this amount of additional capacity is possible, though, unless utilities make substantial new orders in the next few years, it will become very difficult to achieve. A detailed, country-by-country, breakdown of the Moderate Growth scenario is also given in Appendix B.

The reader may be interested in comparing several recent alternative growth scenarios, in addition to the two used in this book. In Table 4-2 we compare forecasts made by the International Atomic Energy Agency and by the U.S. Department of Energy with both the Present Plans and Moderate Growth scenarios.

REACTOR REQUIREMENTS FOR URANIUM

With appropriate assumptions about uranium requirements for particular types of reactors, we are now in a position to derive overall uranium needs for the Present Plans and Moderate Growth scenarios. In doing so, we use the reactor requirements and operating assumptions listed in Chapter 1 (Table 1-1). While the tabulations above of reactor capacities do not distinguish between reactor types, the actual computation of uranium needs

Table 4–2.   Comparison of Recent Nuclear Power Growth
Projections (GWE).

	Source			
	IAEA	*U.S. DOE*	*Moderate Growth*	*Present Plans*
Projections for 1985				
France	38.3	25.7–29.1	35.6	43.1
W. Germany	16.5	15.0–16.6	16.4	17.6
Other Europe	37.5	38.0–39.0	39.0	39.9
Total Europe	92.3	78.7–84.7	91	100.6
Japan	26.0	17.8–19.5	21	22.8
Other	25.1	20.1–23.1	23	24.9
Non-U.S. WOCA	143.4	116.6–127.3	135	148.3
United States	80.3	72.3–90.1	91	96.4
Total WOCA	223.7	188.9–217.4	226	244.7
Projections for 1990				
France	59.1	38.5–44.2	54.8	56.3
W. Germany	23.1	22.5–37.5	24.1	26.3
Other Europe	46.6	44.8–52.1	50.1	50.5
Total Europe	128.8	105.8–133.8	129	133.0
Japan	46.0	24.5–30.9	37	25.9
Other	36.9	33.9–44.7	39	37.2
Non-U.S. WOCA	211.7	164.2–209.4	205	196.1
United States	114.0	111.7–121.0	117	121.3
Total WOCA	325.7	275.9–330.4	322	317.4
Projections for 1995				
France	72.0	50.1–56.5	67.2	56.3
W. Germany	28.2	28.7–33.9	31.5	28.8
Other Europe	71.15	56.4–73.0	69.3	51.4
Total Europe	171.35	135.2–163.4	168	136.5
Japan	68.0	40.8–46.2	53	26.2
Other	50.6	51.2–69.3	62	40.1
Non-U.S. WOCA	289.95	227.2–278.9	283	202.8
United States	122.4	113.0–127.3	131	123.9
Total WOCA	412.35	340.2–406.2	414	326.7

Table 4–2. *(continued)*

		Source		
	IAEA	*U.S. DOE*	*Moderate Growth*	*Present Plans*
Projections for 2000				
France	NA	60.7–73.5	77.8	56.3
W. Germany	NA	34.9–45.3	37.7	28.8
Other Europe	NA	78.2–105.8	89.5	51.4
Total Europe	NA	173.8–224.6	205	136.5
Japan	NA	45.1–54.4	68	26.2
Other	54–88	80.9–118.5	89	40.1
Non-U.S. WOCA	NA	299.8–397.5	362	202.8
United States	NA	109.4–139.6	150	126.1
Total WOCA	504–558	409.4–537.1	512	328.9

Footnotes to Table 4-2

- IAEA forecast from: OECD Nuclear Energy Agency and the International Atomic Energy Agency. 1983. Uranium Resources, Production and Demand. Paris, France: Organization for Economic Cooperation and Development.
- U.S. DOE forecast compiled by the Energy Information Administration of the U.S. Department of Energy; non-U.S. forecasts from Andrew W. Reynolds, "The Outlook for Nuclear Power," Uranium Colloquium V, Grand Junction, Colorado, October 1982; U.S. forecasts from R. Diedrich, "Estimates of Future U.S. Nuclear Power Growth," SR–NAFD-83-01, January 1983 Draft.
- Moderate Growth scenario is adapted from the forecast from the May 1982 Nukem Market Report (NUKEM GmbH, Hanau, West Germany).
- Present Plans scenario based on the utility survey conducted by *Nuclear News* (publication of the American Nuclear Society), 1983 February. Survey includes only reactors for which there is a firm utility order and a specified date of operation.

does take into account the differing requirements of particular reactor types. Our assumption in making these calculations is that enrichment plants operate at a tails assay of 0.20 percent and that reactors operate at a capacity factor of 70 percent, above the average for experience to date.

Figure 4–2 shows the results of this calculation for the two reactor scenarios for nuclear power outside the United States excluding the centrally planned nations (non-U.S. WOCA). We have also computed historic requirements for reactors that have been put into service since the beginnings of commercial nuclear power. The latter calculation is somewhat idealized: no one knows exactly how much uranium has been used worldwide. However, the estimate here is probably not far wrong. In early years, fuel burnups were lower than at present, but reactor capacities were relatively small,

Figure 4-2. Reactor Uranium Requirements under Alternative Growth Scenarios.

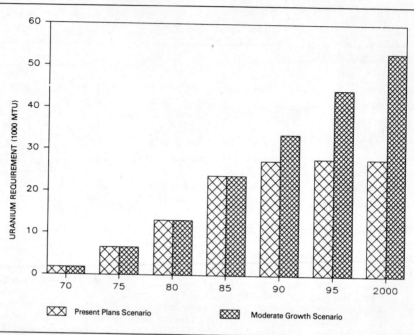

Source: Author's calculations based on growth scenarios of Tables B-4 and B-5, Appendix B, utilizing fuel-cycle assumptions of Table 1-1.

leading to only a small error in the direction of underestimating uranium used. More recently, capacity factors have been lower than the 70 percent assumed, leading perhaps to a slight overestimate of actual uranium consumption. Under these assumptions, reactor consumption of uranium outside the United States totalled about 127,000 MTU through 1982. A similar calculation shows U.S. consumption of about 101,000 MTU over this same period.

EFFECTS OF ENRICHMENT CONTRACTING

As discussed in Chapter 1, enrichment supply arrangements—both contracts and equity commitments—have a major impact on uranium demand. Many such arrangements were made at a time when nuclear power growth was expected to be more rapid than it now appears. As a result, the imperative to deliver uranium in order to satisfy enrichment feed requirements is greater than would be justified simply by reactor requirements for uranium. In Chapter 1, we presented a compilation of contractual and equity commitments to enrichment supply from the sources discussed there. The data on which this table is based are the most recent publicly available—from early 1983. However, it should be noted that resales and adjustments in contracts and in enrichment plant operations have occurred since these numbers were compiled, and thus they should be regarded only as portraying the general dimensions of the enrichment contracting situation.

It is useful to compare these enrichment commitments with reactor requirements for enrichment implied by the Present Plans and Moderate Growth scenarios. Again using the fuel-cycle parameters of Chapter 1, and taking account of differences between reactor types, we have computed these needs for the two scenarios. The results for non-U.S. users are presented in Figure 4-3 where they are compared with the enrichment-supply arrangements detailed in Chapter 1. As is evident, past and existing future supply arrangements are considerably in excess of reactor needs under either scenario not only during the past decade but also until about 1989.

After 1989, consumer commitments to enrichment supply decline rapidly, implying a need to make new commitments. The magnitude of such commitments—and the source of supply—appear to depend critically on whether new nuclear power plants are ordered. In Figure 4-3, the curve shown as "available supply" includes present commitments to supply from the United States, full utilization of Eurodif capacity, growth in Urenco utilized capacity to 4 million SWU annually, and reliance on the Soviet Union for no more than 3.2 million SWU annually. If no new reactors are ordered, this supply will be more than adequate for the remainder of the century. That is, there is little need of new foreign commitments to supply from the United States, or the Soviet Union, if utilities use the appreciable inventories built up over nearly two decades. It is only if significant new reactor capacity is added—in addition to that on order—that new supply must be sought. Such new supply might come from Urenco expansion, construction of new enrichment capacity in Japan or Australia or elsewhere, or new supply agreements with the United States.

Excess enrichment supply commitments imply a need for uranium that may exceed reactor requirements for those reactors utilizing enriched uranium. Thus, enrichment commitments tend to drive uranium demand until 1987 or 1988 under either nuclear growth scenario. This will result in a continued buildup in enriched uranium inventories unless successful efforts are made to reduce overall enrichment commitments; simply reallocating enrichment services will not change this basic fact.

Figure 4-3. Enrichment Supply, Contracts, and Reactor Requirements (*Non-U.S. WOCA*).

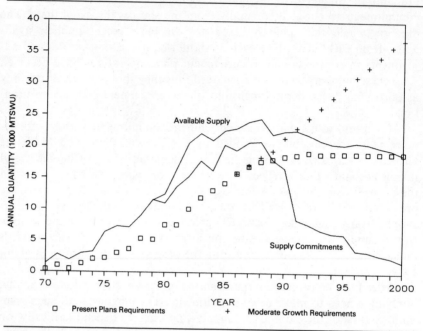

Source: Enrichment deliveries from Figure 1-5. Does not include 10,020 MTSWU advance sale by the United States to Japan in 1973. Reactor requirements for enrichment are author's calculations based on historic nuclear power expansion, fuel-cycle requirements of Table 1-1, and the nuclear growth scenarios of Appendix B.

Note: Available supply assumes no further purchases from the United States, expansion of Urenco annual capacity to only 4,000 MTSWU, and a maximum of 3,200 MTSWU per year from the Soviet Union. Much larger quantities of enrichment services will actually be available.

The implications of this picture are not only that substantial stocks of enriched material already exist but that they will continue to build up over the decade unless there is a significant reduction in enrichment deliveries. There will also be a significant overcapacity in enrichment until perhaps the end of the century. Enrichment customers will be under continuing economic pressures to reduce enrichment commitments where there is the flexibility to do so. It also suggests that suppliers of enrichment services may experience difficulty in justifying planned capacity expansion, or even full operation of existing capacity.

Because uranium prices have declined significantly since their high point in 1978, while enrichment prices have not declined, the optimal tails assay for enrichment has increased for consumers able to purchase at or near today's uranium prices; that is, it would be better for customers to use fewer units of enrichment and deliver more uranium in order to produce the same amount of fuel. This departure from economic optimality means that there will be downward pressure on enrichment prices: at some point it becomes rational for customers to accept cancellation penalties, to resell SWUs at a discount, or to risk pressuring enrichment suppliers to alter contract terms.

INVENTORY DEMAND

During much of the past decade, many utilities did not explicitly regard nuclear fuel inventories as a major planning issue. Radidly rising prices for uranium, and security of supply concerns, encouraged commitments beyond those necessary just to fuel reactors. As a result, inventories grew steadily, if not at the same rate everywhere. With the decline in uranium prices and the reduction in security concerns, however, many utilities have begun to consider inventories as an important factor in planning. This importance has been enhanced by financial problems for many utilities: excess inventories represent either a continuing cost, or a sales opportunity that might improve utility balance sheets in an era of reduced growth and high interest rates.

Thus, where before there might have been a positive demand for inventories, this demand in many cases is now, in effect, negative, with sales of inventories displacing primary demand rather than adding to it. The result is a secondary market in both uranium and enrichment. This secondary market adds to what is already a great excess of capacity and primary market supply to put additional downward pressure on prices. The extent to which these secondary market sales have pushed prices below the levels that would otherwise obtain is uncertain, though the effect is clearly important. We shall return to the discussion of secondary markets—and to a rough calculation of inventory levels—in Chapters 8 and 9, after a detailed review of conditions in both producer and consumer nations.

TECHNOLOGICAL CHANGE IN
THE FUEL CYCLE

Many of the advanced fuel-cycle research and development and demonstration programs in the world were motivated by a belief that conventional nuclear fuels would soon become scarce, insecure, and increasingly high in price. Technological efforts have proceeded on several fronts. Major efforts have been made to improve the efficiency of use of fissile material in conventional or modified light-water reactors. Recycle of plutonium in light-water reactors, breeder reactors, and fusion have been pursued largely as a way to reduce the demand, moderate the prices, or alleviate the security concerns associated with conventional low-enriched uranium fuels.

Recent changes in nuclear fuel markets—most importantly, the surplus supply and weak prices, as well as high stock levels—tend to reduce fears about future scarcity, insecurity, and high prices. As a result, they tend to reduce the urgency for technological change and for advanced nuclear technologies, as these relate to nuclear fuel supply. While in the long-term there may still be concerns about uranium and enrichment supply, the point in time at which this concern becomes pressing has been put off further into the future, well into the next century.

A case in point is the slow speed with which commercial reprocessing is proceeding. In the United States and in West Germany, there have been major delays and still no firm plan or policy. The Barnwell facility in the United States is virtually complete, but despite positive views of reprocessing by the Reagan administration, the administration's desire that the private sector be responsible has thus far not encouraged private firms in the United States to accept the responsibility for operating the plant (except perhaps under government guarantees). West Germany's small (40 MTU annually) demonstration reprocessing facility at Karlsruhe is being reconditioned. Plans to build a new facility of at least 350 MTU annual spent-fuel capacity at Gorleben were frustrated by political difficulties, though such a facility may be built by the mid-1990s. In the interim, DWK (Deutsche Gessellschaft für Wiederaufarbeitung von Kernbrennstoffen) has planned away-from-reactor, spent-fuel storage facilities (totalling 3,000 MTU of spent fuel) at Gorleben and Ahaus. These storage facilities may be in operation by 1985.

Japan's Tokai Mura facility has experienced not only significant political problems (raised by U.S. nonproliferation concerns) but also technical difficulties. The "second" reprocessing plant is not planned to be in operation until sometime in the 1990s. In the interim, spent fuel from Japan and European nations is being sent to France or the United Kingdom.

Until 1981, the French program was the most aggressive in the world, with plans to expand rapidly to a reprocessing capacity of 2,000 MTU annually. A major debate in late 1981 appeared at first likely to stall the program (and interrupted shipments from Germany and other nations). But the program was reaffirmed, perhaps in part because of the extensive contracts held with foreign utilities (Belgium—400 MTU, West Germany—2,140 MTU, Sweden—670 MTU, Switzerland—470 MTU, Japan—2,200 MTU, and the Netherlands—120 MTU).[7] These contracts—totalling about 6,000 MTU of foreign spent fuel—put most of the cost burden and other risks on the foreign utilities. (France even has the right to specify the nature of the waste products to be returned to the country of origin.) For those utilities that have dealt with France on reprocessing, the contracts have ensured that wastes will be taken care of by others for perhaps a decade or more (though at a high cost), but the long-term waste problem has only been deferred and changed in form, not entirely solved.

In the United Kingdom, British Nuclear Fuels, Ltd. (BNFL) has plans for an oxide fuel reprocessing plant to start up in 1990, with capacity growing eventually to 1,200 MTU annually. As in France, spent fuel—which includes about 4,400 MTU of foreign light-water reactor fuel under reprocessing contract with BNFL—will be stored in pools until it can be reprocessed.

A recent survey[7] estimates the reprocessing capacity expansion worldwide as shown in Figure 4-4. Also shown is the possible range of uranium used in reactors worldwide (derived from our estimates of reactor fuel use under our two different growth scenarios). What is of greatest importance here is the observation that even on a year-by-year basis (and ignoring the backlog of spent fuel), reprocessing capacity even in 1995 is unlikely to be much more than 10 percent of the annual conventional uranium use of all the utilities outside the United States (non-U.S. WOCA). Indeed, if more reactors are ordered (our Moderate Growth scenario), the percentage decreases to only a little more than 6 percent. When the backlog of spent

Figure 4-4.   Planned Reprocessing Capacity.

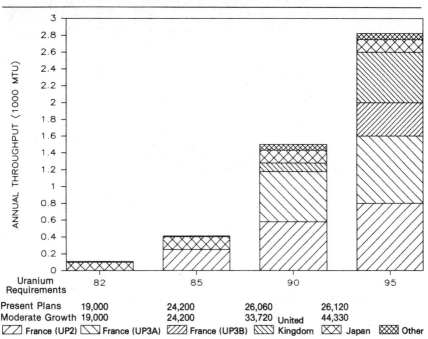

Source: Nukem Market Report. April 1982. Hanau, West Germany: NUKEM GmbH.

fuel is taken into account, it is evident that currently planned reprocessing capability will affect only a very small percentage of fuel throughput.

The plutonium and recovered uranium resulting from this reprocessing program would have only a small impact on the demand for conventional low-enriched, light-water reactor fuel. The fissile plutonium and unburned U-235 in one kilogram of spent reactor fuel is enough—under ideal conditions—to replace about 2.2 kilograms of natural uranium feed that would otherwise be enriched and used in light-water reactors. Thus, in 1995, the total amount of spent fuel reprocessed in planned facilities would yield enough fissile material to replace at most 6,000 MTU annually of natural uranium demand. Of course, the lead time for such recycling may be five or more years so that this uranium savings would not be realized until later. In total, the uranium and plutonium recovered by reprocessing from 1982 through the year 1995 would replace only about 2.3 percent of the uranium that would have to be purchased through the year 2000 under the Moderate Growth scenario (under the Present Plans scenario, the percentage savings would be greater, but there would also be less demand for the material because of lower uranium prices). Recycled plutonium and uranium material (or comparable amounts of U-235 from natural uranium) can provide greater energy output if used in advanced thermal reactors or in breeder reactors. But this is also true if U-235 from natural uranium is used as the fissile material in advanced reactors or the breeder. Thus the fissile atoms in natural uranium will always have a value comparable to those in manmade fuels such as plutonium.

Whether such savings are important will depend on uranium market developments in the first half of the next century. It has long been recognized that conventional fission technologies will have a strong cost advantage over new and more capital-intensive technologies (such as the breeder or fusion) until uranium prices increase so much as to justify the additional investment. At that point, the old and new nuclear technologies will be in equilibrium—with some small-scale penetration by the new technologies—until there is a clear cost advantage. It is not that the new technologies will undermine the competitiveness of today's light-water reactors and the value of uranium, but rather that the costs of this now traditional technology and fuel will have to rise, or be expected to rise, until investments in new technologies are justified.

REFERENCES

1. For a review of uranium as a commodity, see: Radetzki, Marian. 1981. *Uranium: A Strategic Source of Energy.* New York: St. Martin's Press.
2. For a review of these technical possibilities, see: "Uranium Resource Utilization Improvements in the Once-Through PWR Fuel Cycle." April 1980. Prepared

for the U.S. Department of Energy by Combustion Engineering Power Systems. Report CEND-380/UC-78.

3. OECD Nuclear Energy Agency, and the International Atomic Energy Agency. 1969. *Uranium Production and Short-Term Demand;* OECD, and IAEA. 1970, 1973, 1976, 1977, 1979, 1982. *Uranium: Resources, Production and Demand.* Paris, France: Organization for Economic Cooperation and Development.

4. *Nuclear News.* Surveys published August 1975, August 1977, August 1979, February 1981, February 1982, and February 1983. La Grange Park, Illinois: American Nuclear Society. Only reactors in operation, under construction, or on order with a startup date specified are included.

5. *Nuclear News.* February 1983. World List of Nuclear Power Plants.

6. *Nukem Market Report.* May 1982. Hanau, West Germany: NUKEM GmbH.

7. *Nukem Market Report.* April 1982. Hanau, West Germany: NUKEM GmbH.

PART II
The Suppliers

5 AUSTRALIA AND CANADA

Canada and Australia seem destined to be the chief sources of low-cost uranium during at least the remainder of this century. Both nations provided uranium for the nuclear-weapons programs of the United States and the United Kingdom until such procurements ended in 1967. Subsequently, however, the uranium industries in these two nations followed somewhat different paths. In Canada, a high level of weapons-related output was followed by a collapse of much of the domestic industry; in Australia, a lower level of weapons-related development meant there was less failure to be experienced. In both nations, government stockpile programs helped cushion the fall.

As commercial demand began to appear in the late 1960s and early 1970s, Canada was much better positioned to take advantage of market opportunities than was Australia. Mining operations expanded rapidly, and Canada profited from the great runup in prices. In contrast, Australia missed most of the boom, in part because of government changes that put a freeze on sales and industry expansion just at the critical moment. Eventually, two major ventures found financing just before prices again fell.

In both Canada and Australia, governments have played major roles in domestic uranium activity and in setting conditions on exports. And in both, similar political and economic factors have interacted to determine policy. Internationally, both nations have been responsive to concerns about nuclear-weapons proliferation, both participated in the uranium cartel that sought to increase returns on natural resource exports, and both instituted national review procedures for exports. Domestically, both have experienced conflicts between central governments and increasingly strong state or provincial interests, and both have acted to limit and regulate foreign ownership of extractive industries.

111

Despite these similarities, however, the ways in which uranium industries have developed have been different in the two nations. And while both nations in recent years have yielded to economic pressures for development and exports, there are significant differences in the market roles of the two nations, differences that may have major impacts on the future international uranium market. In the following sections, we explore conditions in each nation in depth.

AUSTRALIA

Australia contains some of the largest and richest deposits of uranium in the world. With slower than expected growth in world demand for uranium and a general oversupply situation, Australia's relatively low-cost supply may play a critical role in the evolution of international nuclear fuel markets. However, the extent and nature of this role will depend on investment decisions, the domestic policy context in Australia as it relates to exports, and on actions taken by other major actors in the international market—most notably the United States. In the following sections we first explore the historical role of Australia in international uranium trade and then evaluate the present capacity and contract status of key ventures. We next review the present Australian policy environment. Finally, we consider Australia's role in the future world uranium market.

Uranium exploration began intensively in Australia in the mid-1940s following a request by the United Kingdom to assist in providing uranium for the weapons programs of the United Kingdom and the United States. This exploration was soon successful, with finds first at Mount Painter and Radium Hill in South Australia and then, in 1949, at Rum Jungle in the Northern Territory. In 1954, the Mary Kathleen deposit was discovered in Queensland. The Australian government offered rewards for discoveries, paying out A$50,000 each to the discoverers of the Rum Jungle and Mary Kathleen deposits. The development of these deposits was financed by the Combined Development Agency—the Anglo-American defense purchasing agency—on a cost-plus basis, or by the U.K. Atomic Energy Authority (UKAEA). Libral incentives were offered, including tax-free rewards for discoveries, guaranteed prices for uranium, and tax exemptions on profits for the companies involved.[1]

The result of these efforts was a rapid increase in uranium output, beginning in 1954 at Radium Hill, and increasing as Mary Kathleen came on line in 1956. Production reached a peak of 1,200 MTU annually in 1961. However, as was the case in Canada and elsewhere, this mineral boom faltered by 1964 with the expiration of agreements with the Combined Development Agency and the UKAEA. Mining ceased at Mary Kathleen and Rum Jungle in 1963, but milling of ore continued at Rum Jungle until

1971 under the Australian Atomic Energy Commission. This operation resulted in a government stockpile of about 1,730 MTU.[2] Over the period 1954–64, Australia exported about 6,050 MTU in connection with activities of the Combined Development Agency and the UKAEA.

In 1967, the Minister for National Development implemented a commercial export policy. This, together with rapidly rising expectations worldwide for the growth of nuclear power, led to a new wave of exploration in Australia. In 1970, major new discoveries—Ranger, Nabarlek, and Koongarra—were made in the Alligator Rivers area of the Northern Territory. In 1972, a large sedimentary deposit was discovered at Yeelirrie in Western Australia. And, in 1973, the Jabiluka deposit was discovered north of Ranger. These discoveries increased the proven reserves of Australia perhaps thirtyfold, ultimately adding more than 300,000 MTU to known resources in just a few years. This rapid expansion in reserves was paralleled by the entry of Australia into the new world market for uranium in 1972, with three companies committing more than 9,000 MTU to foreign utilities for delivery over the period 1976–86.[1,2]

However, the election of a Labor government in December 1972 was accompanied by a reexamination of uranium development and export issues. A hold was put on new export approvals for Mary Kathleen, Ranger, and Nabarlek. While existing contracts were honored, the policy questioning effectively restrained further development and sales activities. In the fall of 1974, the government announced a new policy, emphasizing governmental participation in future uranium activity (including mine investment and export authorities). Its key conditions specified that:

- Greater government ownership and financing would be exercised through the Australian Atomic Energy Commission (AAEC), beginning with a 50 percent share in Ranger and financing of nearly three-quarters of the development costs;
- Sole authority to explore—beyond existing licenses—in the Northern Territory would be vested in the AAEC;
- Existing contracts would be honored, using the AAEC stockpile and output from Mary Kathleen production and, eventually, from Ranger; and
- All future sales would be made by the AAEC.

In the summer of 1975, the Labor government, under the terms of the new Environmental Protection Act of 1974, initiated a public inquiry into proposals for development of uranium deposits in the Northern Territory. This commission, chaired by Justice Fox, resulted in two reports, in 1976 and 1977. While the commission raised numerous issues that needed to be resolved in connection with uranium mining and international sales, these issues were not considered cause to rule out such activities by the new Liberal-National Party Coalition government led by Malcolm Fraser to whom the report was delivered. The Fraser government acted[3] on May 24, 1977, to institute a new policy oriented toward private investment and export. The reelection of a Labor government in 1983 has again raised questions about the nature of future uranium development in Australia. We shall return to a detailed discussion of this policy context following a review of the uranium industry and trade patterns.

Industry Structure

The uranium industry in Australia expanded rapidly when the new government policy toward uranium development was announced in 1977. Three major new deposits—Ranger and Nabarlek in the Alligator Rivers region of the Northern Territory, and Yeelirrie in Western Australia—received development approval. Subsequently, but prior to the election of the Labor government, Commonwealth development approval was granted to the small Lake Way deposit in Western Australia, the massive Jabiluka deposit in the Northern Territory, and the Honeymoon deposit in South Australia. The state government of South Australia also approved development of the immense deposit of uranium, copper, gold, and associated minerals at Roxby Downs. The locations of these existing and prospective mining developments are shown in Figure 5–1.

Historically, approximately 9,000 MTU of exports had been contracted for by 1972, before a change in national government resulted in a pause in most uranium development and export activity. Of these early contracts, about 4,000 MTU was to come from Mary Kathleen, about 2,500 MTU from Ranger, and about 2,500 MTU from Nabarlek.[2] Deliveries of this material began in 1977, with some borrowing from the government stockpile necessary in early years; these early contracts all expire by 1986.

Contracting began again in 1979, and since then more than 40,000 MTU of additional material has been committed, either under contract or through equity participants. In the following we review each of the approved and possible projects in greater detail. Export commitments are author's estimates, based on a large number of (frequently conflicting) public sources and private discussions.

Mary Kathleen.   Until mid-1980, Mary Kathleen in Queensland was the only commercial uranium mine operating in Australia; it has also been the first to shut down. With lower world prices, relatively high costs, and limited reserves—and thus little opportunity for further sales—Mary Kathleen Uranium (MKU) stopped production in the fall of 1982, when enough material had been produced to meet existing contracts. Rehabilita-

Figure 5-1.   Uranium in Australia.

OECD Nuclear Energy Agency and the International Atomic Energy Agency. 1982. *Uranium Resources, Production and Demand.* Paris, France: Organization for Economic Cooperation and Development, p. 206.

tion of the mine area is expected to take about two years. MKU's 1972 contracts included a sale to Commonwealth Edison of the United States (1,630 MTU), two sales to Kernkraftwerk Brunsbüttel of West Germany (1,240 MTU), and separate contracts with Tokyo Electric (770 MTU) and Chugoku Electric (380 MTU) of Japan. The quantities given here are approximate—small adjustments in quantities and delivery schedules have occurred.

The ownership of Mary Kathleen Uranium Ltd. was shared by the Australian Atomic Energy Commission (41.6 percent), Conzinc Riotinto Australia now CRA Ltd. (51 percent), and the remainder by private Australian investors.[4] CRA Ltd. is owned 57.2 percent by Rio Tinto Zinc

(RTZ) of London.[5] RTZ had owned a higher percentage (72.6), but reduced its share in order to come closer to Australian equity requirements. Prices for MKU uranium were set on an annual basis through negotiation, but were related to prevailing market prices. As a consequence of low ore grades, high costs, low world prices, indebtedness, and union opposition that has delayed deliveries, MKU has sometimes operated at a loss.

Ranger.   The facilities presently in operation at Ranger process ore from the first (number 1) ore body which contains about 48,000 MTU at a grade of 0.25 percent yellowcake. A second ore body (referred to as number 3) reportedly contains more than 36,000 MTU, at a slightly lower grade than that of the first. Other estimates put total resources at more than 100,000 MTU. Additional reserves may be added through an active drilling program.

The Ranger deposits were originally discovered and initially developed by a joint venture of Peko-Wallsend and Electrolytic Zinc Industries. In 1974, the Labor government took a 50 percent share in exchange for development financing. In August 1979, the Commonwealth government announced that it would divest itself of its 50 percent share. The result was the organization of a new company, Energy Resources of Australia Ltd. (ERA), led by Peko-Wallsend Ltd. The Australian government reportedly received more than $A100 million for its share. The present ownership structure is:[6]

	Percent
Peko-Wallsend	30.5
EZ Industries	30.5
JAURD (Japan)	10.0
West German group	14.0
OKG (Sweden)	1.0
Australian Public	14.0

JAURD—the Japan Australia Uranium Resources Development Company—is a consortium of Japanese utilities and a trading house (Kansai Electric with 5.0% of Ranger, Kyushu Electric with 2.5%, Shikoku Electric with 1.5%, and C. Itoh and Co. with 1.0%).[7] The West German group involves subsidiaries of the utility RWE (Rheinishe Braunkohlen, 6.25% of Ranger), Urangesellshaft (UG Australia, 4.0%), and Saarberg-Interplan (Inter-Uranium Australia, 3.75%).

Project loans to finance Ranger totalling up to US$390 million were reportedly arranged from JAURD (capitalized at $18.18 million and with a loan from Japan's Export-Import Bank of $140 million)[8] and a bank consortium led by J. Henry Schroder Wagg and Company Ltd. and the Continental Illinois National Bank and Trust Company of Chicago.[6] This amount was about equal to actual development costs, and thus most of the financing for Ranger came from foreign sources, including the United States.

The public shares of Ranger (14% of ERA, totalling 57.5 million shares) are widely held by 37,117 investors, half also being shareholders in Peko-Wallsend and EZ Industries.[9,10] The offering price of the original stock was A$1.00. By the end of November 1980, book value of the company was A$1.26 billion, thus more than doubling in a short time and making it among the top ten Australian companies. Assuming 85,000 MTU of reserves, the original share price translated into a value of A$2.20 per pound of uranium oxide reserves.

All of Ranger's output at the initial production level is under contract. Two contracts date from 1972: one with Chubu Electric of Japan for 1,000 MTU, and one with Kyushu Electric for 1,540 MTU, both originally to be delivered between 1977 and 1986. Further sales were made subsequent to the formation of ERA: 1,920 MTU for delivery to Korea Electric between 1983 and 1992; 1,730 MTU for delivery to Indiana and Michigan Electric (United States) between 1982 and 1990; 2,420 MTU for delivery to Sweden's OKG (also an equity participant) between 1982 and 1996; and 1,200 MTU for delivery to Belgium's Synatom in the late 1980s. In 1983, two contracts were reported signed with U.S. utilities—Virginia Electric Power Co. and Wisconsin Electric Power Co.—although neither has been submitted for government approval as of this writing. The total amount involved in these two contracts appears to be about 300 MTU per yr. Most of these sales were arranged directly with customers rather than through agents.

Much larger quantities will go to the members of the German and Japanese consortia that have taken equity positions in ERA. More than 15,000 MTU will go to the three German companies over the period 1982–1996, while about 11,000 MTU will go to the Japanese members of JAURD. The latter quantities are much larger than equity percentages would indicate. It should be noted that some of the participants in both the German and Japanese groups act as agents; it is therefore not entirely certain that all of the uranium involved will ultimately go to Japan and West Germany (though we assume this in our projections of supply available to consumer nations).

Yellowcake production began at Ranger in September 1981, though official startup was not expected until November. A first shipment of more than 230 MTU was made in October,[11] indicating that there was no difficulty in reaching, or perhaps surpassing, design output. More than 860 MTU was borrowed from the Australian government stockpile for preoperational deliveries to Chubu and Kyushu Electric, and it will have to be repaid from future Ranger production. With the above delivery schedule, the output of the Ranger project, at the initial 2,540 MTU annual level planned, is fully committed through the 1980s and nearly three-quarters committed even into the mid-1990s. While exact delivery schedules are not public (and indeed are subject to change due to labor difficulties and other problems), it is possible to construct an approximate schedule that is consistent with the known facts. This schedule is reflected in Figure 5-2, later in this chapter.[7]

Output for the first nine months of operation (October 1981 through June 1982) was 1,977 MTU. An undisclosed fraction of this was returned to the Commonwealth government to repay loans from the stockpile used to meet delivery commitments prior to startup; income from this transaction (in which ERA essentially was able to "sell" at the higher prices prevailing in 1977-1979) was reported as A$28.9 million. Income from sales to foreign concerns was reported as A$117.1 million. Profits were reported as A$37.9 million. Deliveries to overseas equity holders were reported as being at about A$38 per pound of yellowcake.[12] Other reports suggest that Japanese utilities are paying more than A$38 per pound in 1983, but whether there is any difference between equity holders and contact buyers is not known.[13] With consumer purchases tied in some cases to mine financing, it is difficult to isolate true transaction prices.

ERA has now committed sufficient uranium to consider expansion of mine and mill capacity (if sales can be made in today's market), an option that has long been included in project planning. In fact, the new Labor government has given ERA permission to seek new sales in the United States and appears to be seeking ways to develop contract terms that would make ERA competitive in the U.S. market. A doubling of mine and mill capacity is possible but would require earlier use of material from the second ore body. The cost of this expansion has been estimated at A$60 million, far below (perhaps one-tenth) the cost of new production capacity at new deposits in Australia (and perhaps anywhere).[14] There is some evidence that ERA's expansion proposal is intended to deter development of new mines that might compete for what market remains.

The production costs for Ranger are not known with precision, but the size of the deposit, the ease of open-pit mining and uranium extraction, and the economies of scale possible, suggest low unit production costs. In addition, "royalties" paid to Aborigines (indirectly, through the Commonwealth government) are small on a per pound basis. Working in the opposite direction are relatively high Australian labor costs; expenses arising from the remote location of the mine; opposition by some unions (necessitating higher transportation expenditures); and possibly higher future royalties to the Northern Territory. Australian government officials indicate privately that actual costs (both capital and operating costs) have turned out to be significantly higher than originally projected.

Nabarlek.   This small but rich deposit was discovered in 1970 in the East Alligator River area of the Northern Territory by Queensland Mines, an Australian company. The range or ore grades is extreme, varying from a fraction of a percent to 72 percent uranium oxide;[15] demonstrated reserves

Figure 5-2.   Australia's Commercial Uranium Exports (*Estimated*).

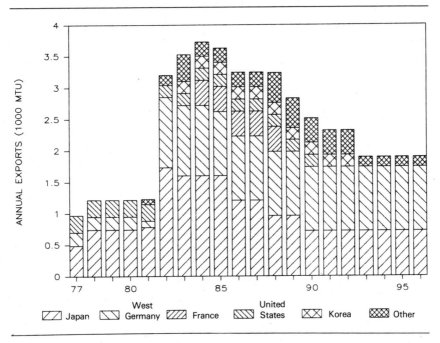

are about 10,000 MTU at an average ore grade of about 2 percent yellow-cake.[16]

Development of the Nabarlek project began in the spring of 1979, following the signing of an agreement with the Northern Land Council (representing Aborigine landholders) and the receipt of a special mining lease from the Northern Territory government. Under the terms of the agreement with the Northern Land Council (NLC), Queensland Mines must pay a 4.5 percent royalty (indirectly, through the Commonwealth government) to the NLC; it must also remit a 1.25 percent royalty to the Australian government (which in turn pays a similar amount to the Northern Territory government). These royalty agreements are subject to review in five years, though proposals were made in 1982 to increase payments to the Northern Territory before that date.

Following these agreements, the entire ore body at Nabarlek was removed and stockpiled for processing over a period of six to ten years. The high grade of the ore has required special precautions in mining and milling, but is also ensures low production costs since the capital investment per unit of output is small compared to many other mines. The mill, which has a rated capacity of about 1,270 MTU per year,[6] began operation in mid-1980 but has consistently produced at levels above this, reportedly at about 1,570 MTU on an annual basis (approximately 850 MTU was produced in the last half of 1980).

Queensland Mines has written four contracts for deliveries of uranium from Nabarlek. Two of these date from 1972: one with Shikoku Electric and another with Kyushu Electric. The total amount committed under these contracts was 3,230 STU_3O_8 (2,485 MTU); deliveries prior to the beginning of Nabarlek production were made from the Australian national stockpile and totalled about 740 MTU. Development of the mine and mill complex was partially financed with A$75 million from Shikoku and Kyushu (75% of which came from Shikoku and 25% from Kyushu).[9] Adjustment in delivery commitments may have occurred in connection with this financing, and there are reports of delays in shipments while Australia and Japan completed negotiations on a new bilateral agreement. Recent reports indicate a total of 4,615 MTU going to the two utilities through 1989.

In 1980, the Finnish utility, Teollisuuden Voima Oy Industrins Kraft-AB (TVO), contracted with Queensland for 690 MTU (900 STU_3O_8) to be delivered over nine years beginning in 1981. The fourth known contract—signed in 1981—is with Electricité de France for 2,000 MTU. These contracts reportedly involve a floor price escalated from January 1980; the nature and implications of this constraint are discussed in greater detail below. The contracts currently held by Queensland account for over 70 percent of total expected lifetime production from Nabarlek. Efforts to sell the remainder—especially in the United States—appear to be encountering difficulty due to market conditions and the export price constraint. Because of the unique mine and mill circumstances, Nabarlek will be producing at levels above annual export delivery commitments and will stockpile material against future deliveries.

In 1983, Queensland's contract with Electricité de France became a focus of conflict between Australia and France over the latter's nuclear-weapons testing in the South Pacific. Although deliveries were not scheduled to begin until late 1984, Queensland sought to make early shipments in mid-1983. Robert Hawke, the new Labor Party prime minister, announced that shipments to France would be embargoed until October 1984 unless weapons tests were stopped (or, as suggested, moved to metropolitan France).[17] Since France has an ample supply of nuclear fuel without Australian deliveries, the Australian threat seems more a debating point than a potentially serious disruption of supply.

The startup of Nabarlek was important to Australia's efforts to meet delivery commitments since continued borrowing from the government stockpile to fulfill contracts for both Ranger and Nabarlek would otherwise have depleted that stockpile in 1980. The higher than expected production at Nabarlek ensured replacement at an early date (sometime in 1981) of the 740 MTU borrowed by Queensland. In the first several years of Nabarlek's operation, there were recurrent difficulties with the Oenpelli Aborigines who objected to transport of mill output along the road to the port of Darwin, despite the agreement with the Northern Land Council that was intended to settle such difficulties. There was also opposition by some transportation unions to handling uranium from Nabarlek. In 1981, Queensland resorted to shipping by air from the mine to Darwin.[18]

Other Ventures—Northern Territory.   There is a great deal of uranium activity still underway in many parts of Australia, despite reduced market expectations. Exploration resumed in the Northern Territory (following a prohibition instituted there in 1975) and in virtually all of the States. A number of ventures demonstrated significant deposits, and several—Yeelirrie, Lake Way, Honeymoon, Roxby Downs, and Jabiluka—received Commonwealth and State government clearances to proceed to seek financing and export contracts. It is not possible here to review all of this activity; instead, we consider only those ventures at the most advanced stage, or most likely to proceed to actual uranium production. We first consider those in the Northern Territory, which contains perhaps the richest deposits. New uranium discoveries are also possible there, and indeed seem likely if exploration proceeds.

Jabiluka.   Discovered in 1971 in the East Alligator River region, the Jabiluka deposit consists of two ore bodies: the first containing about 2,800 MTU at an average yellowcake concentration of about 0.24 percent, the second containing about 170,000 MTU at an average concentration of about 0.39 percent. Jabiluka thus contains the largest quantity of uranium at commercially attractive concentrations thus far delineated in Australia (or elsewhere, for that matter). The deposit is owned by Pancontinental Mining Ltd. of Australia (65%) and the U.S.-based Getty Oil Development Company Ltd. (35%).[16] This allocation does not meet the strictest equity requirements of the Commonwealth government. Approval by the Liberal-National Country government was contingent upon Getty reducing its share to 25 percent prior to exports of uranium from Jabiluka.

Planning at the large Jabiluka deposit was far advanced prior to national elections in 1983. Agreement with the Northern Land Council on payments to Aboriginal groups had been reached, Commonwealth approval was granted in July of 1982, and a mineral lease by the Northern Territory government was approved in August 1982. However, Jabiluka has thus far been unable to find commercial support to proceed. Major marketing efforts have been underway, but success in the current market, and with Australian price restrictions discussed later, is unlikely. Discussions with Britain's Central Electricity Generating Board (CEGB) about delivery of 1,650 MTU over a ten-year period have not led to an explicit contract with the CEGB, or precipitated a wave of new purchase commitments.

Changing market conditions and the need to secure various approvals led to some changes in development plans for Jabiluka. Originally, both ore bodies were to have been mined by open-pit methods. To reduce environmental impacts, it was decided to pursue an underground mining approach. This is inherently more expensive, though there will be a considerable reduction in the amount of overburden material that will have to be moved, resulting in a partially compensating savings. Planned mill size has been increased to about 3,800 MTU annually. While expansion to more than 7,600 MTU annual capacity had been planned for about five years after startup, expansion would obviously depend on market conditions.[19] At its initial level, Jabiluka would have an output comparable to that of Rössing in Namibia, the largest uranium mine in the world. The new attitude toward expansion was probably related to the process of getting approvals in Australia: firm plans to expand to a level that would undoubtedly have a major impact on the depressed world market would increase the stakes in the approval decision.

Whether and when Jabiluka goes ahead will depend largely on sales opportunities, which in turn depend on Australia's national export policies, especially on price provisions. The current government-specified floor price, which is above spot prices and likely to remain so for some years as the general oversupply situation works its way out, makes it difficult for the owners of Jabiluka to secure customers for its large volume of output. And without customers, there will not be the capital to finance such an expensive venture—estimates for which range from A$600–800 million. Jabiluka must also compete for customers with ERA at Ranger, which might increase output by as much as the total prospective output of Jabiluka at far lower cost. In 1982, those associated with the project proposed a startup in 1987; subsequent developments suggest that this can no longer be considered.

Koongarra.   The Koongarra deposit is a relatively small deposit (resource estimates range from about 11,000 to 25,000 MTU), but its high grade (with average grades reported variously as between 0.35% and 0.70% yellowcake) is comparatively close to the surface. It could thus be mined by open-pit methods at costs that would be low compared to many other mines outside Australia. However, the mining site—an area 2.5 sq km—is located wholly within the proposed Kakadu National Park (Ranger and Jabiluka are on the boundary of the initial park area) and is upstream of ecologically sensitive areas. Special legislation removed the mine area from the park in 1979, though approval of Aborigine groups has continued to be a major issue. In 1983, the new Labor Party government's Minister for Aboriginal

Affairs asked the Northern Land Council, representing the Aborigines, to stop negotiations with the mining company until the new government formulated a policy on uranium development. However, the NLC decided to initial an agreement in July rather than delay significantly. Such an agreement might form the basis of a request for compensation to Northern Territory groups, should the government prevent development of the deposit.

Until 1980, rights to the Koongarra deposit were owned by Noranda Australia Ltd., a wholly owned subsidiary of Noranda Mines of Canada. Subsequent to filing of a draft environmental impact statement in 1979, however, rights to the deposit were sold to Denison Mines, also of Canada, reportedly for about C$120 million (or about $4 per pound of reserves). Initial plans were for a mine and mill with an annual capacity of about 1,000 MTU to begin operation in the late 1980s. However, as part of the government approval process, Denison would have to divest itself of up to 75 percent of its equity in the deposit to Australian concerns, or seek exemption from this requirement. While the policy of the new Labor government is not yet clear, early indications are that it will be unwilling to allow development to proceed.

Venture in Other States.   A number of other deposits have been identified in the Northern Territory and others are likely to be found. However, those already identified are generally smaller and lower in grade than Ranger, Jabiluka, and other deposits that are already being pursued actively. In addition to Northern Territory activities, attention focuses on prospects in other States of the Commonwealth. Several deposits have been identified in Queensland, the most attractive of which is Ben Lomond. In Western Australia, Yeelirrie and the Lake Way deposit are evident candidates. And in South Australia, three deposits—Beverley, Honeymoon, and Roxby Downs—have been considered for development.

Yeelirrie.   The third major new deposit to receive development approval under the Liberal-National Country Party government was that at Yeelirrie Station in Western Australia. The ore is deposited in calcretes in fossil valleys in the granitoid gneiss in a zone nearly 9 kilometers long, and consists of about 39 million tonnes ore at an average grade of 0.14 percent yellowcake (using a cutoff of 0.05%). The deposit thus contains about 42,000 MTU, though at an ore grade somewhat lower than that of Ranger. Because of the character of the ore, the Yeelirrie project is proceeding in two stages. The first stage, initiated in 1979 after approvals were granted by the Commonwealth and State governments, involved operation of a relatively small pilot plant to evaluate extraction technology and costs, further delineation of the ore body, and evaluation of possible water supplies. The pilot plant began operation in late 1980 and the first-stage operations concluded at the end of 1982, at a cost of about A$24 million. A decision on the second stage—the engineering and construction of a mine and mill and the necessary infrastructure—was scheduled to be made in 1983. As at Jabiluka, the biggest obstacles will be winning approval from the new Labor government and securing financing. It is likely that financing will only be available if mine proponents can exhibit either long-term contractual obligations with customers at prices considerably higher than those now prevailing in the spot market, or major financing commitments by an equity holder. Approval by the new government will be even more problematic.

Originally, the Yeelirrie project was a joint venture between Western Mining Corporation (through its wholly owned subsidiary, the Yeelirrie Development Company) with a 75 percent equity, Esso Exploration and Production Australia, Inc. with a 15 percent equity, and Urangesellshaft Australia Ltd. with a 10 percent equity.[16] Yeelirrie thus satisfied the most rigorous Australian equity criterion. Esso had agreed to pay 80 percent of development costs and take 50 percent of the output but in 1982 announced plans to sell its share. Western Mining, which was to market 40 percent of output, is reportedly talking to several foreign companies, including Electricité de France (EdF), about taking over Esso's role in the project.[20] However, EdF—and presumably other potential partners—are unlikely to be happy with the terms being offered. Unless someone can be found to replace Esso as an equity partner or major long-term customer, this mine is unlikely to proceed, even if government approvals are given.

Ben Lomond (Queensland).   Reserves at this deposit appear small by Australian standards, perhaps only four or five thousand tonnes of uranium, though some reports put the grade at about 0.32 percent. The deposit also contains molybdenum that could be coproduced. Rights are owned by Minatome Australia Ltd., a wholly owned subsidiary of Minatome of France. Evaluation of the deposit is still at the feasibility stage and little information has been released. Minatome is reported to be considering a mine and mill complex capable of producing about 4,000 tonnes of uranium and 2,500 tonnes of molybdenum over a ten-year period. An environmental impact statement was prepared but not submitted to the government, since both the uranium and molybdenum markets are depressed.

Lake Way (Western Australia).   This deposit, discovered in 1977 about 70 kilometers north of Yeelirrie, is reported to contain about 3,400 MTU at an average yellowcake concentration below 0.1 percent. While this is not a large nor particularly rich deposit, it may be possible to mine it economically since it consists of a single seam that is easily accessible from the surface. The environmental impact statement submitted to the government of Western Australia puts development cost at A$45 million: at a hypothetical annual production rate of 450 MTU, the capital component of the costs would be well below A$10 per pound.[21]

Rights to the Lake Way deposit are now owned by Vam Ltd. of Australia (46.5%) and Delhi International Oil (53.5%). Both Vam and Delhi increased equity holdings when Wyoming Mineral (a Westinghouse subsidiary) sold its equity in 1981. Wyoming was to receive about 20 MTU of annual output as part of the transaction. Delhi was acquired by CSR Ltd. of Australia in 1981, making it possible for Lake Way to meet Commonwealth equity requirements for development.[22] Development approval was granted by the Fraser government early in 1982; however, development seems unlikely to proceed under the new Labor government.

Beverley (South Australia). This uraninite deposit is in sandstone lenses containing about 11,500 MTU and is located at a depth of about 100 meters, of which about 9,800 MTU is regarded as recoverable. The grade averages 0.24 percent. Rights to the Beverley deposits are owned by Western Nuclear Australia, a subsidiary of Phelps Dodge of the United States (which owns 50%), Oilmin (16.66%), Petromin (16.66%), and Transoil (16.66%). The final three are all Australian companies.[16]

Original plans for an open-pit mine were changed in favor of *in situ* solution mining. Under this plan, production capacity would be added in increments of about 200 MTU of annual capacity, beginning in 1986 and expanding in output until 1992. Under the new plan, an estimated A$515 million would be spent over the expected twenty-nine-year life of the venture (about A$20 per pound of recovered yellowcake in current dollars). The uranium concentration at Beverley is relatively high, and the formation encourages efficient extraction. Western was willing to provide financing for a pilot plant as early as 1982 if approval were obtained. Under the Fraser government, development would have depended on gaining approval of an environmental impact statement and compliance with national equity requirements. However, the new Labor Party government in South Australia has said it will not approve the development.

Honeymoon (South Australia). This deposit contains between 2,000 and 3,000 MTU. Rights are owned by Mount Isa Mines Ltd. (49%), AAR Ltd., a CSR subsidiary (25.5%), and Teton Australia Pty. Ltd. Co., a subsidiary of UNC Resources, the large U.S. producer, (25.5%).[6] Plans to recover about 100 MTU annually (in the initial phase) by solution mining were well advanced when general approval was granted by the Commonwealth government in 1982, conditional on changes in equity positions to meet national requirements. However, the new Labor government in South Australia refused a production license for the mine early in 1983, granting instead a retention lease that would only allow the owners to preserve their interests in the deposit. The partners, who say they are legally unable even to continue evaluation of solution mining at the site, have asked the government for compensation.

Roxby Downs (South Australia).   This large deposit at Roxby Downs Station was discovered late in 1977 by Western Mining Corporation as a result of geologic and geophysical evaluations beginning in 1975. Located more than 350 meters below the surface and at places several hundred meters thick, the deposit has no surface manifestations but may contain the largest ore body of uranium—and one of the largest copper deposits—ever found.[12,23] In the 18 square kilometers drilled by mid-1980, no firm boundary had yet been found; similar mineralization—another deposit or, perhaps, an extension—has been found 25 kilometers from Roxby Downs.[6] While estimates of resources are difficult under such conditions, the environmental impact statement filed in mid-1982 put uranium reserves in an area 4 kilometers by 7 kilometers at more than 1 million MTU; copper reserves in this area were put at more than 32 million tonnes and gold at 1,200 tonnes.

Drilling indicates that copper may grade at an average concentration of 1.6 percent and uranium at about 0.06 percent. Gold and various rare earth minerals are also present. The ratio of copper to uranium concentrations is about 25 to 1; the ratio of prices of copper and uranium (both of which are currently depressed) is about 1 to 30. Thus copper and uranium might be coproduced on a roughly equal footing.

In 1979, Western Mining formed a joint venture with BP Australia Ltd. (a subsidiary of British Petroleum). Under this agreement, BP's role is largely financial, with project management in Western Mining's hands. BP will provide A$50 million for exploration and other feasibility studies (which includes a major exploratory shaft) and will finance development of a mine and mill with a production capacity up to 150,000 tonnes per year of copper, 2,500 MTU annually of uranium, 3,400 kilograms of gold, and 23,000 kilograms of silver annually. By mid-1982, A$80 million had been spent, with planned expenditures of A$150 million through the end of the feasibility stage.[14]

Development of the Roxby Downs deposit seems ultimately inevitable, though there is uncertainty about when this might be economically attractive or politically feasible. The ratio of waste to recoverable minerals is quite high, entailing relatively high costs, though scale economies may reduce unit costs somewhat. Given that copper and uranium are about of equal importance (and that the copper market may take some years to recover from its current oversupply) and that gold, silver, and other coproduced products will add to recovered values, it is useful to compare this deposit with uranium deposits of more than twice the grade (say, 0.15%) but with similar overburden. Such a comparison suggests that costs per pound of recovered uranium may be higher than some other deposits in Australia, perhaps comparable to the average mine in the United States. It is probably therefore correct to view Roxby Downs as a potentially major contributor to the uranium market in the 1990s or after the year 2000; such a time scale also seems consistent with the scope of the project envisioned.

While the political climate a decade or more hence is uncertain, there now seem few obstacles to development of the Roxby Downs deposit. The earlier election of a Liberal government in South Australia resulted in the lifting of the prior Labor government's restriction on new uranium exports. But the restoration of a Labor government in November 1982 has not altered the positive governmental outlook for Roxby Downs, in part due to the 6,000 or more permanent jobs expected to be created in the economically depressed State. The State government approved the environmental impact statement in mid-1983, and construction of a pilot plant was to begin immediately, with completion scheduled for the end of 1984. The joint venture agreement between Western Mining and BP was approved by the Commonwealth government, suggesting that BP will not be subject to the most stringent form of the domestic equity requirement.

Export Commitments and Production Plans

Overall, Australian production capacity now in place—from Nabarlek and Ranger—is quite well matched to export commitments, as shown in Figure 5-2. The delivery commitments shown should be regarded as approximate, in part because precise delivery schedules have not been made public but also because changes in schedule due to events in Australia and in consumer nations are possible. Not shown are borrowings and repayments to the Australian national stockpile of uranium. What is important here is the general level of contract commitments compared to mine and mill expansion.

Under current arrangements, Japan is by far the largest customer for Australian uranium, despite its relatively strong commitments in other markets. The next largest recipient is West Germany, whose relatively recent contracts with Australia contribute to a new core of long-term coverage of uranium needs (see Chapter 8). France is a relatively new entrant to the Australian market, diversifying supply away from African producers. As yet, the United States is but a minor customer for Australian uranium.

Three tentative supply arrangements were pending in mid-1983. The Central Electricity Generating Board of Britain was said to have agreed to take 1,650 MTU over a ten-year period from Jabiluka, though there are indications that this is not a firm commitment. Queensland is to be actively seeking buyers for its surplus production from Nabarlek.

As Figure 5-2 shows, deliveries to date plus current contract commitments through 1990 are for about 30,000 MTU, or about 86 percent of actual and prospective production of about 34,500 MTU through 1990 for the two mines now operating. A large fraction of the additional output that might be available will come from Nabarlek. It should be noted that due to high grades and a small mill investment, Nabarlek has considerable flexibility in changing output levels, or in stockpiling for potential future deliveries.

It is evident from the preceding section that very substantial expansions of Australian production capacity are technically feasible. If all of the ventures listed above were to go ahead by 1990, production capacity could exceed 20,000 MTU annually, a capacity that would make Australia the dominant supplier to the world market. This is, of course, unlikely to happen. Development thus far has been dependent on securing long-term contracts, often in combination with substantial foreign financing. Given the current oversupply situation and, especially, a shift toward spot-market procurement strategies by utilities, it is unlikely that the critically-needed, large-scale financing will be available to new Australian ventures until it is again possible to write long-term supply agreements. The most likely near-term development is expansion of Ranger—where marginal costs are low. Given the variety of minerals to be produced at Roxby Downs and the political commitment to proceeding, development there seems ultimately guaranteed, though the timing is uncertain.

Policy Context

As discussed briefly earlier in this chapter, Australian uranium policy has undergone a complex series of changes since 1970. However, many foreign observers have overstated the significance of such shifts, sometimes confusing rhetoric and domestic political movements with fundamental policy shifts. Although the two major Australian political parties differ significantly in their attitudes toward uranium mining, the regulatory environment under which mining and exports occur does not seem likely to change radically. The ways in which statutes established under the Fraser government, described below, may be applied may be changed by subsequent governments but will probably continue to govern the industry's operations for the foreseeable future.

The policy announced by the new Liberal-National Country Party government in 1977 was more positive in its attitude toward uranium development and exports than that of the preceding Labor government. The policy statement of 1977 dealt with nonproliferation safeguards, the question of whether mines should be developed sequentially, Aborigine land rights, radioactive waste, mining codes, further environmental standards, controls and supervision, the creation of a national park in the Northern Territory, the role of the government in uranium marketing, and taxation of resources.

Subsequent to the 1977 policy statement, a number of actions were taken by the Fraser government that implemented, clarified, or supplemented that statement. These included the negotiation of a host of bilateral agreements governing nuclear commerce with other nations, studies of possible enrichment and conversion activities in Australia, a reallocation of State and Commonwealth responsibilities for regulation, the establishment of a Uranium Export Office and a policy on how sales of uranium are to be negotiated, a statement of conditions for prior approval on retransfer and reprocessing, implementation of policy on foreign equity participation in uranium development, proposals for new royalty structures, and others. Approval was subsequently granted for development of a number of new deposits. Many new export commitments had been made from some of these approved projects and approval for several new ventures was pending at the time of the 1983 Labor election victory.

The Liberal-National Country Party policy, as implemented, was clearly responsive to major economic and political forces present in Australia, and in its relationships with other nations. These include: the nature of (changing) relationships between the Commonwealth government on the one hand and the States and the Northern Territory on the other; the desire to realize the greatest possible value from resources on the part of a nation whose exports are dominated by minerals and agricultural products; a balancing need to attract foreign capital and trade opportunities in order to develop natural resources; the need to build and sustain a viable nonproliferation regime while accommodating the legitimate nuclear power objectives of Australia's allies; the imperative of sustaining political leadership despite intense political differences among groups; the necessity of accommodating aboriginal land, custom, and economic interests; and the need to maintain good relationships with allies—such as Japan and the United States—that happen also to be consumer nations, or to have their own uranium and other nuclear industries. All of these pose similar challenges to the Labor government elected in 1983. In the following sub-sections we examine each of the major policy areas that remain relevant to Australia.

Safeguards. Under the 1977 policy, exports were to be made only to nuclear-weapons states that do not use Australian-origin uranium for explosive or military purposes and to non-weapons states that are parties to the Non-Proliferation Treaty. All recipients must allow International Atomic Energy Agency (IAEA) safeguards on uranium of Australian origin. Trade will take place only under bilateral agreements that ensure use for peaceful purposes, physical security measures, fall-back safeguards (in case IAEA safeguards should fail to apply), and prior consent to retransfer, enrichment to greater than 20 percent U-235, and reprocessing. Specification of the way in which this prior consent would be exercised was reserved in the 1977 policy statement and in early bilateral agreements, pending the conclusion of the International Nuclear Fuel Cycle Evaluation initiated by the Carter Administration.

Between 1978 and 1981 bilateral safeguard agreements were concluded with Canada, Finland, France, the Philippines, South Korea, Sweden, the United States, the United Kingdom, and with Euratom. The Euratom agreement is of particular interest since, unlike parallel efforts by the United States and Canada, it successfully dealt with the requirements of the European Economic Community (EEC) agreement for free movement of fuel between member states. In the Australian bilateral agreements with individual European nations, there are no difficulties with transfers to other nations also having bilateral agreements with Australia. Transfers to other nations not having bilaterals with Australia for purposes of processing (as, presumably, for enrichment in the Soviet Union) are also allowed, with the consumer nation providing assurance that the material (or its equivalent) will be returned to it. There is thus an important mutual element of trust—or, perhaps, delegation of responsibility—in this approach.

Negotiations with Japan proved more difficult than those with European customers: Japanese utilities have contracts with European reprocessors and Japan wanted—in effect—to be treated as if it were a member of Euratom, with free movements between Japan and Europe allowed under the proposed bilateral agreement with Australia. It was not possible to reach agreement on this issue until early in 1982, when an agreement was initialed by both nations and subsequently ratified. In the interim, some shipments to Japan under new contracts were held up. Agreement with Switzerland has also been slow in coming, with an agreement initialed in 1983 but not ratified before the change in Australian government.

An interesting, if somewhat peripheral, difficulty in implementing Australian safeguards arose from the fact (identified in the Fox Commission report) that IAEA safeguards become fully effective only at the point of conversion of uranium oxide to uranium hexafluoride. Since this presently occurs outside Australia, there would have been a putative gap in control over Australian uranium. Initially it was thought that this could be overcome by maintaining Australian title to material until it had entered the conversion step and attracted rigorous safeguards. However, the result of U.S. antitrust laws and various litigation (relating to cartel allegations) would then have exposed the companies involved to the possibility of seizure in connection with such legal actions.[24,25] It was subsequently argued that Australian requirements for physical security, notification of movement of material, intergovernmental arrangements with processor countries, and bilateral agreements with ultimate purchasers are adequate. Why this was not simply asserted in the first place is not clear.

Prior Consent.   A critical issue in these and other bilateral agreements has been the manner in which Australia would exercise the prior-consent right over retransfer and reprocessing embodied in its 1977 policy. In 1977, Prime Minister Fraser had indicated that Australia would reserve its position on reprocessing pending the outcome of the International Nuclear Fuel Cycle Evaluation (INFCE) and other international studies.

In November 1980, following the completion of INFCE, the government announced the procedures that would be adopted in making decisions about whether or not it should give its prior consent to reprocessing of Australian-origin material for the nuclear fuel-cycle programs of countries with which Australia has nuclear safeguards agreements. A statement of this policy was made by the Minister for Foreign Affairs in the House of Representatives:

> The first step will be the provision by an interested customer country of detailed information. The purpose of this will be to establish the country's need, within its overall energy strategy, for Australian origin nuclear material to be reprocessed. I should like to emphasize that this initial information gathering stage in no way seeks to intrude into the decisionmaking process of negotiating partners and that information received would remain confidential between the parties to respect commercial confidences. . . .
>
> If the need to reprocess Australian origin nuclear material is demonstrated to the satisfaction of the Government and the controls and safeguards applied to program involving reprocessing meet all the Government's existing requirements, the Government will be prepared to consider granting its consent to the reprocessing of Australian origin nuclear material in specifically defined nuclear fuel cycle programs on the following bases:
> - agreement in advance to reprocessing for the purposes of energy use;
> - agreement in advance to reprocessing for the purpose of the management of materials (plutonium, fission products and unused uranium) contained in spent nuclear fuel;
> - case by case consideration of requests for consent to reprocessing for other peaceful non-explosive purposes including research;
> - storage and use of plutonium of Australian origin separated from spent fuel to be in ways that do not cause proliferation dangers;
> - provision for consultation and review of the operation of the agreed conditions; and
> - commitment by customer countries to support the development of more effective international control measures relevant to reprocessing, including an international plutonium storage scheme.[26]

In the debate over nonproliferation conditions led by the Carter administration (as discussed in Chapter 2), there was always a tension between the need to provide assurance to civilian nuclear programs and the desire to control—if not eliminate—"proliferation-sensitive" activities such as reprocessing and recycle. The latter desire seems to have been based on an underlying belief that safeguards were not adequate to deal with such sensitive activities. The Australian policy clearly came down on the side of assurance, asserting that safeguards can be made adequate, and moreover that a freer Australian export policy in fact contributes to nonproliferation goals. This is clear in the continuation of the Minister's presentation:

> The Government believes that advance consent to reprocessing of Australian origin and nuclear material within specific programs designed for energy use and management of materials contained in spent nuclear fuel is a legitimate procedure which can be achieved consistently with nonproliferation objectives. Customer countries have an understandable interest in wishing to know, when they purchase Australian uranium, for what purposes they may use it. The predictability and stability introduced by advance consent for certain specified uses, combined with commitments to stringent safeguards and controls, represent a positive contribution to non-proliferation objectives and stable international trade in the nuclear field.[26]

With less aggressive policy leadership on the nonproliferation issue under the Reagan administration in the United States, and with a more general recognition that fuel-cycle leverage does not easily compel the behavior of allies (as contrasted with nations less trustworthy), the more flexible Australian policy is unlikely to encounter opposition abroad. Indeed, the primary judges will be the Australian public, whose consciousness has been raised on this issue. Whether the present political consensus on nonproliferation policy will endure depends in part on external events—such as further proliferation events—that might alter public perceptions of Australian responsibility.

Environmental Constraints and Aboriginal Interests.   Environmental protection and the protection of the interests of aboriginal peoples are key social objectives of the new uranium policy and are implemented through both Commonwealth laws (the principal of which is the Environment Protection—Impact of Proposals—Act of 1974) and laws and regulations promulgated by the States and the Northern Territory. Under the Australian federal system, the States retain substantial regulatory powers; the present trend seems to be toward even greater State presence in local regulation, permission processes, and other local aspects of development. As it progresses toward greater independent political identity, the Northern Territory has also taken on a larger role.

The process of achieving environmental clearance for uranium projects—through consideration of an environmental impact statement—therefore is primarily a State (or Northern Territory) function, though the same process also satisfies Commonwealth requirements. Exceptions to this have to do with obligations that are perceived largely at the national level: the development of the Kakadu National Park (which abuts or includes several uranium prospects and contains a particularly sensitive environment), aboriginal rights, and national and international economic and political interests. The Commonwealth has the ultimate power to deny development approval to projects that fail to meet these nationally perceived domestic needs, or that fail to meet international policy constraints.

The process by which aboriginal interests are protected is through negotiation between potential developers and a body representing the aboriginal groups—the Northern Land Council. The council appears to have been generally successful thus far in securing royalties for the aboriginals, protection of sacred areas, employment opportunities, and other social welfare objectives, though a few tribal groups have continued to act independently, refusing to accept agreements made on their behalf. The most serious of these has been the reluctance of the Oenpelli to allow transport of uranium from Nabarlek along the road to the port of Darwin.

Foreign Participation.   Exploration is generally open to foreign interests, though the government "expects foreign interests to seek Australian participation in those projects that can reasonably be expected to proceed to the development stage."[27] While most mining projects require only 50 percent local equity, uranium is deemed of such importance that a higher equity participation is desired. A recent statement of this policy is given in government documents:

> The Government expects new projects for the mining and production of uranium to have 75 percent Australian equity and to be Australian controlled. In cases where 75 percent Australian equity is clearly unobtainable, alternative proposals will be considered. In such cases, it would need to be demonstrated satisfactorily that:
>
> (a)   75 percent Australian equity is unavailable;
> (b)   the project would be of significant economic benefit to Australia;
> (c)   there would be at least 50 percent Australian equity; and
> (d)   Australian participants would have the major role in determining the policy of the project.
>
> Where projects do not have 75 percent Australian equity and Australian control, arrangements may be required to increase Australian participation over an agreed period.[27]

Of all the ventures already approved or pending approval, only Ranger, Nabarlek, and Yeelirrie satisfy the most stringent equity requirements. It appears likely that some exceptions will be made, under the conditions listed above, for several of the projects still awaiting full approval. Getty's prospective large capital investment at Jabiluka and BP's at Olympic Dam may make foreign equity holdings greater than 25 percent attractive to the Australian government. In recent years, between 50 and 60 percent of capital investment has come from outside Australia. While Australia finds it desirable to finance development as much as possible from domestic sources, foreign investment plays a critical role in the Australian economy and in Australia's balance of trade.

Development Approval.   As noted above, the Commonwealth has authority to grant or deny approval to new mining ventures on the basis of whether they meet national and international objectives. In principle, this power could be used to control the rate of uranium development in Australia, thus restraining ventures that would lead to undue price competition and excess supply, or that might trigger protectionist efforts in the U.S. market. The Fox Commission had proposed sequential development of uranium projects (at least for Northern Territory deposits), but this formal approach was rejected by the Fraser government in its 1977 policy statement with the observation that sequential development would be the natural course of events simply because of the different rates of progress of various ventures.[2] Of course, this denial of the formal sequential development approach still preserves fully the power of the Commonwealth government to influence the rate of expansion of Australian uranium production capacity.

Development approval also depends—in the Australian federal system—on the States. As noted above, the States have substantial regulatory power in environmental and other areas. The basic laws governing the Commonwealth's role in nuclear fuel-cycle activities are the Atomic Energy Act of 1953 (as amended), the Customs Act of 1901, and various environmental laws and regulations. This legal basis has been under reexamination with the expressed intention of possible revisions. Such revisions would be undertaken in consultation with, and take into account the particular interests of, the State and Northern Territory governments and would "be established as far as possible under State and Northern Territory legislation."[25]

This shift reflects current trends in federal-state relationships,[28] trends that are likely to continue independent of which party is in power. The Commonwealth government generally has the power to collect taxes and other revenues, but many social and other programs are conducted by State governments. The result is that revenues must flow back to the States to fund programs over which Canberra has primarily only fiscal control. And over the past decade the volume of such transfers has increased, with the federal government bearing the political costs of raising the funds. This, taken with the fact that political organization is on a State basis (with a concomitant importance attached to issues of relevance to particular States), suggests a rationalization of responsibilities, with the States and the Northern Territory taking greater control, and perhaps fiscal responsibility, for developments occurring within their geographic areas. Such changes would also give scope to the varying interests of different States in uranium development and nuclear fuel-cycle activities more generally, without the need to force a national consensus.

In the nuclear area, expanded State roles might include safety and other regulation, and investment in infrastructure associated with nuclear fuel-cycle activities, including enrichment facilities as well as uranium mines and mills. It might also mean royalties and taxes to be collected locally to meet these increased responsibilities. In such a situation, the Commonwealth would still retain responsibility for overall national economic interests (including pricing) and for foreign relations (including nonproliferation) but would leave much of the rest to the States. This issue remains somewhat speculative, but the evolution of federal-state relationships in nuclear fuel-cycle matters will certainly influence Australia's future role in the international market.

Conditions for Export. In addition to its power over development approval, the Commonwealth government must also approve conditions for exports. Once a company has development approval, it may enter into sales agreements with foreign consumers and agents. But prior to entering into such negotiations, it must consult with the Minister for Trade to determine in advance the minimum terms and conditions for contracts that will satisfy the government's ultimate requirements for exports. These terms include:

- terms and conditions consistent with the government's nuclear safeguards policy;
- the duration of the contract and the quantity of uranium to be sold;
- the use to which the uranium is to be put by the purchaser;
- the price to be charged.[24]

The Australian government argues that this prior determination is necessary to avoid exposure to U.S. antitrust laws which—in matters involving U.S. commerce—operate even extra-territorially.[25] By imposing constraints prior to contract negotiations, the Australian government frees the participants in those negotiations of exposure to such antitrust laws.

In addition to the antitrust justification, Australian officials also advance arguments concerning national economic and political interest at stake in the export control process:

> Broadly, the Federal Government looks to resources development to achieve economic growth and to bring substantial and realistic benefits to the Australian community. It relies on its powers to regulate the export of key commodities to ensure that the national interest is preserved.
>
> One objective of this control is to ensure that Australia receives world market prices for its exports and fair and reasonable contract conditions. It must be made clear in this regard that in the Government's view, there is no room for a "cost plus" approach to pricing—an approach which would inevitably lead to an inefficient industry and to the view that justifiably expected returns are being realised elsewhere. Either could be expected to lead to resentment and hostility within Australia.
>
> . . . The export control powers also underpin the Government's policy of project development approval and its safeguards policies. They ensure observance of the policies essential to all the public acceptance of uranium mining.

They are effective because all uranium production is for export, Australia having no domestic demand.[24]

The prior-determination process gives the government significant control over Australia's role in international market developments, including the pricing of uranium. It also gives it continuing power—beyond the development-approval stage—over the rate of expansion of domestic production capacity since many ventures will not proceed without contracts with buyers, or without strong assurance that it will be possible to write such contracts. While these powers certainly exist, their use is publicly seen as relating to "normalizing" markets rather than manipulating them:

It should be emphasized, however, that Federal Government oversight is not intended to hamper normal commercial negotiations. The Australian Government is firmly committed to the principle of private enterprise development of the nation's resources. The use of export controls in this context is to neutralise market imperfections rather than to dictate terms. Interference in marketing dynamics would not be in the long term interest of a vigorous and competitive industry or of the Australian companies which must remain competitive in world markets.[24]

The question of Australia's relationship to the world market will be returned to below.

Finally, the federal government has a third level of control over exports at the point of individual shipments under approved contracts from approved mining projects. Such shipments require a Restricted Goods Export Permit that may be issued under the Customs (Prohibited Exports) Regulations if:

- the uranium is being exported from a project that has the status of Government development approval;
- the export is for the purpose of performing an approved contract; and
- the Australian safeguards policy is fully complied with.[24]

Taxes. Uranium mining ventures operate under the general federal tax system (46 percent on profits) and pay various taxes and fees to local governmental bodies: to the States, the Northern Territory, or to aboriginal land holders. At present, the latter taxes and fees are generally quite small.

However, both taxes and other fees may be increased in the future. In its decision of 1977, the Commonwealth government explicitly indicated the possibility of a future resource-based tax on earnings from uranium development.[29] Subsequently, both States and the Northern Territory government have proposed increases in taxes and other impositions. Initial reactions from industry have been understandably negative. But the economic rewards from uranium mining—given low production costs and prices held up by a federal price floor on exports—have naturally created a competition for the large economic rents available. It can be expected that most of the parties to this competition—the companies, the States and the Northern Territory, aboriginal groups, unions, the Commonwealth government, and others—will seek their share of the benefits.

Among political jurisdictions, the Commonwealth may well accede to royalties or other payments to local governments if those governments take larger independent fiscal responsibilities for local needs, including the investments in infrastructure associated with uranium activities, and responsibility for resolving the local political problems arising from such activities. The more difficult problem may arise in satisfying other interest groups. Here there is incentive for at least some of these groups—unions, Aborigines, or others—to use obstructionist tactics to secure a larger share of the income from ventures that are allowed to go ahead. Such efforts may tend to delay project approval and development, or interfere later with the performance of export obligations. Both types of events have been evident.

Processing and Value Added.   There is a natural desire on the part of any nation with extensive natural resources to realize as much as possible the economic opportunities associated with the processing and upgrading of those resources. In the case of uranium, such opportunities arise in connection with conversion to UF_6, and enrichment to levels suitable for use in light-water reactors. Australia at present possesses neither capability. However, discussions are underway with nations able to transfer these technologies to Australia.

With the encouragement of the previous Commonwealth government, a private joint venture, the Uranium Enrichment Group of Australia (UEGA), was established in 1979 to consider the commercial feasibility of a uranium enrichment industry in Australia. This series of studies culminated in 1982 with the selection of centrifuge technology and association with the European Urenco consortium. The approach taken strongly suggested that any such enrichment venture would be a private endeavor, with the government taking responsibility only for nonproliferation and safeguards. The approach that might be taken by the new national Labor government is not clear yet. Given the great excess supply of enrichment services in the world market, commercial opportunities would at best arise only near the end of this decade, or sometime in the next, depending on the course of nuclear power development.

In the case of uranium conversion—a less demanding technology than enrichment—the government of South Australia in 1981 entered into a joint venture with British Nuclear Fuels Ltd. and two private companies to evaluate the feasibility of a uranium hexafluoride conversion plant in the Port Pirie area of South Australia.

A central question (among both consumer nations and other suppliers) has been the extent to which purchasers of Australian uranium would be required to utilize local enrichment and conversion facilities as a condition of export. Such a condition was apparently included in one of the first uranium supply contracts written after Australia's return to the world market in 1977 (that with South Korea). Reportedly, that clause was not rigorous in its restrictions, including conditions that would have allowed South Korea to purchase services elsewhere if Australian prices were not competitive. Australian officials report that subsequent uranium contracts have not included requirements for utilization of Australian services. The lack of a guaranteed market for domestic enrichers and converters will mean a higher level of market risk for those ventures. One consequence was a preference for centrifuge enrichment technology because of the shorter lead times associated (which allow greater time to resolve uncertainties about demand before a commitment is made) and the ability to expand capacity in small increments.

Political Differences and Union Opposition.   During the tenure of the Fraser administration, the policies of the Liberal-National Country Party coalition were opposed by some trade unions and by some members of the Labor Party. While union opposition was far from unified, there were local instances of refusal to handle or transport uranium. Deliveries from all of the mines now exporting uranium were delayed by union objections on occasion. Such problems did not threaten exports altogether—since alternative transportation arrangements were usually possible—but they did increase costs. In the longer term, one must ask whether there is a basis for more cohesive union opposition to exports. Such a concern is recurrent in uranium industry discussions. However, it should be realized that uranium production and exports benefit important labor groups, and increased revenues will create important constituencies supporting continued exports. Perhaps recognizing this late in 1981, the Council of Trade Unions "temporarily" lifted its ban on handling of uranium exports.[30]

Perhaps more serious are the possible continuance of philosophical differences between at least some members of both parties. One difference concerns the role of the government and the public in uranium decisions and in regard to the allocation of revenue and other benefits. The Labor government of the 1970s participated directly in projects (in investments and, potentially, in profits), and safety and nonproliferation concerns seemed to dominate other factors in many decisions. In contrast, the Fraser government favored private sector initiative and sold off government shares in all ventures except Mary Kathleen (for whose higher-cost operations a buyer could not be found). It also sought to decentralize regulation and some decisions among the States and the Northern Territory. And the conclusions of INFCE concerning reprocessing and plutonium use and management (and, perhaps, the election of a new U.S. administration) provided opportunities to embrace an export-oriented policy without high international political costs.

The position of the new Labor government toward the issues is not yet clear and appears not likely to be clarified for some time yet. However, the nature of the opposition response to the policy implemented by the Fraser government, especially that on prior consent, may give some indications of the issues that must be confronted by the new government, either at some leisure—given market conditions—or perhaps more rapidly if world events force Australian decisions. A sampling from the remarks of the Deputy Leader of the Opposition (Mr. Bowen) in the House of Representatives suggests the potential nature of Labor dissent from continuation of the Fraser policy:

> The Labor Party's policy is still clear; it is fundamental and is directed towards the purposes of peace. We would not mine and sell uranium until such time as the world were able to demonstrate that it could be used safely, that it would not be used for nuclear proliferation and that there would be adequate technology to dispose of the high level waste once it had gone through the first cycle use. Those matters have not been solved. . . .

> It is no disadvantage to wait until such time as we are able to develop that technology. Assuming that we reach that stage, would it not be in Australia's interest also to own and control the fuel throughout the cycle if we have developed the technology, instead of leaving us as the poor relation in the world, the one who gives the advantage to somebody else? This applies to our whole mineral resource area. . . .

> [President Carter] knew that if he allowed reprocessing he would escalate the danger to the world. I have no doubt—this is perhaps unfortunate—that President Reagan will abdicate from that responsibility and say that he sees no danger there.

> . . . The whole issue is whether we should be selling at any low price at all. With a real understanding of commercial transactions we know that it will be a priceless fuel because nobody else in the world will have enough of it, certainly by the year 2000.

> What we have done, as a government, is to take the lead in making the world an even more dangerous place. We are about the first government to abdicate consent and control over safeguards of this material. We have given that away. The commitment now is to get the sale.[26]

Much of this is simply political rhetoric. However, the themes here are evident: there is an assertion both that there has been a loss of Australian control over potentially dangerous situations, and that decisions about economic aspects have been made that are contrary to the national interest. Such views may be relevant when considering the policies which the Labor Party will enact now that it is in power, although early expressions of the attitudes of the current leadership—which includes Mr. Bowen as Minister for

Trade—have not been as categorical as those above. Changes in world conditions—such as further proliferation events—could precipitate to a change in this situation. Consumers that experienced the shifts in Canadian and U.S. policies[31] in the late 1970s cannot entirely discount such future possibilities, though both recent governments have made efforts to sustain Australia's image as a reliable supplier.

Labor Party Policies: 1983.   The 1983 election victory of the Australian Labor Party, both nationally and in state governments, has re-opened many questions about Australian uranium policy. However, it seems unlikely either that these questions will be resolved at an early date or that, when confronted, there will be any radical change in policy. In the near term, the government seems likely to be preoccupied with combatting the combination of high inflation and high unemployment that helped bring it to power. This would not be a good time to forego the economic and employment benefits brought by uranium mining. As for the uranium industry, the new government is currently attempting to formulate a coherent policy representing a compromise between the left-wing of the party and the more moderate leaders. It should be noted, however, that much of Australia's export policy—including pricing—has been developed by the executive branch under laws that are very general in their scope. There are thus many different ways in which the details of export policy may be changed by a new government, without any need for changes in legislation.

During the late 1970s, the Labor Party's stance on uranium was quite extreme, calling for mine closings and abrogations of existing contracts. However, the intensity of these positions on the part of the party out of power seems to have been inversely proportional to the distance from elections and were in most cases not formally adopted. At a party conference in July 1982, with elections anticipated by year's end, a concerted effort was made to moderate the platform to avoid what the party leader at the time, Bill Hayden, described as the "devastating" effect the country would suffer as a result of large-scale contract repudiation.[32]

As a result, the party reiterated its opposition to uranium mining on the basis of proliferation concerns and environmental problems, but dedicated itself to a phase-out rather than an abrupt termination of the industry. Although new mines were to be prohibited, provision for approval of ventures in which uranium was "mined incidentally to the mining of other minerals," was made—apparently an escape clause for the development of the massive Roxby Downs deposit.[32]

The result of a compromise, the wording of the pre-election policy statement was somewhat vague. When elections were held in late 1982 in South Australia and nationally in early 1983, the Liberal-National Country Party tried without success to bring the Labor Party's opposition to uranium development to the fore as an issue. Since the elections, however, internal disputes have created problems for Labor Party leaders as policy development has conflicted with both economic realities and ideological positions. At this writing, a cabinet-level study of options available to the government has been conducted but no action has been taken.

The question of the opening of any new mines (except that at Roxby Downs) has arisen at both state and national levels. The new South Australian government refused to grant a production license for the Honeymoon project and informed the owners of the Beverly deposit to expect similar treatment. The national leadership of the new government has generally agreed with the policy of inhibiting new uranium-mining developments. However, the ERA consortium at Ranger has been given permission to seek new sales in the United States, and Western Mining has been told it can proceed at Roxby Downs. Queensland Mines also received permission to seek new sales from Nabarlek, but this would involve unsold production rather than expansion.

Labor government officials have generally portrayed uranium development decisions as motivated largely by economic considerations, rather than by basic political objectives, implying that Labor is not necessarily opposed to such developments per se, though it seems unlikely that major sacrifices of principle will occur, especially if economic and other rewards are not large. In arguing against the smaller deposits like Honeymoon, the Labor Party also cited concerns about nuclear proliferation, but, as pointed out by the deposit's owners, these concerns are equally valid for Roxby Downs. Early appearances—if not well-argued Party policy—thus suggest an effort on the part of the national leadership to pursue a pragmatic approach to uranium development in which investment, economic benefits, and employment play important roles.

While such a pragmatism is evident in what has thus far been done, there is not complete agreement within the Labor Party on the proper course of action. Opposition to new export permission for ERA and Queensland was strong inside the Labor Party, and critics forced a party caucus vote after permission was given. Thus, general domestic and foreign policy goals and the need to accommodate widely divergent views within the party may lead to unpredictable uranium policy movements. One indication of this volatility was the sudden decision to ban uranium exports to France on the eve of Hawke's visit there, rather than putting the issue on the agenda for discussion, that suggested efforts to satisfy constituencies at home and to send a signal to France that had little to do with uranium trade.

In considering the uranium policies adopted by the Fraser government, the critical question is whether the Labor government will make only incremental adjustments or undertake wholesale changes, as was the case in 1972. The former seems the more likely course but with the new government giving increased scrutiny to domestic environmental and aboriginal concerns and taking a harder line on international issues. The latter include nonproliferation safeguards and provisions for treatment of nuclear wastes resulting from use of Australian uranium. From the perspective of foreign buyers, the reconsideration of export conditions could have major implications. Many customer nations only recently signed new bilateral agreements with Australia and would regard unilateral changes as seriously reducing the attractiveness of Australian supplies.

As is the case with domestic uranium policy in Australia, there may also be a tradeoff in the international sphere between economic benefits and ideology. Australia's floor price has guaranteed returns higher than the spot-market prices charged by other suppliers, though such prices have slowed the rate of uranium development in Australia. If the new government seeks to continue this pricing policy *and* moves to tighten up political conditions on exports, it will undoubtedly be choosing to limit further Australia's future role in the international market. Domestically, the Hawke government seems likely to favor going ahead where economic factors are large and to say no to those mining ventures where economic incentives are small. Internationally, matters are more complicated, in part because it is not obvious that there are large new economic benefits in the near-term except perhaps in the U.S. market, where it is possible that customers might be more willing to accept Australian conditions. Because all of this is uncertain, the Hawke government seems likely to defer to major decisions pending further study.

CANADA

Uranium was first produced in Canada as an incidental byproduct of radium and silver mining operations conducted by the Eldorado Gold Mining Company at Port Radium on Great Bear Lake in the Northwest Territories. Between 1933 and 1940, a stock of about 300 metric tonnes of uranium had been accumulated when mining operations were shut down due to poor conditions in the radium market. In 1942, the United States privately requested of the Canadian government that the Port Radium mine be re-opened for recovery of uranium to be used in the U.S weapons program. In 1943, the Canadian government mandated for itself complete control over exploration and mining of radioactive substances, and in 1944, the Canadian government acquired the shares of Eldorado and formed a crown corporation called Eldorado Mining and Refining, Ltd.

Two years later, the Atomic Energy Control Act was promulgated, reserving to the national government—through an Atomic Energy Control Board (AECB)—control over all regulatory matters having to do with atomic energy and nuclear materials. In 1947, the AECB relaxed restrictions on private involvement in uranium exploration and development in the Canada Lands—the Yukon and the Northwest Territories—an action emulated by provincial authorities (who retained *ownership* of mineral resources). By the early 1950s, "cost-plus" incentives (averaging about U.S. $25 per kilogram of uranium) offered by the U.S. and the U.K. weapons programs had induced a massive, and successful, exploration effort, resulting in production of uranium by Eldorado and by several private companies. Similar incentives were being offered in Australia. But in Canada the scale was greater, as were the difficulties when weapons demand decreased rapidly in the mid-1960s.

Canadian production reached a peak of 12,200 MTU in 1959, a factor of ten larger than the 1961 Australian peak.[33] Production during this period came from eleven mills in the Elliot Lake Area and three near Bancroft (both in Ontario), from two mills in the Northwest Territories, and three in northern Saskatchewan. However, it was also in 1959 that the U.S. Atomic Energy Commission announced that it would not contract further for Canadian uranium. Deliveries under existing contracts, scheduled to expire in 1962 and 1963, were stretched out, some until 1966. An existing agreement with the United Kingdom's Atomic Energy Authority was also renegotiated to extend from 1963 to 1971, but at a much lower price (about five dollars a pound).

By 1967, Canadian output had dropped to 2,800 MTU as a result of termination of U.S. and U.K. military procurement contracts and a protectionist embargo on imports by the United States (U.S. production dropped from a peak of 14,500 MTU in 1960 to about 7,500 MTU in 1966). The effect on Canadian uranium export earnings was even more severe, as shown in Figure 5-3. In 1959, uranium export revenues peaked at nearly C$312 million, making uranium second only to petroleum in resource exports that year. However, by 1967 uranium earnings had fallen to less than C$24 million. As is evident in Figure 5-3, both the great runup in revenues and the subsequent decline were directly traceable to purchases by the United States. During the mineral boom, the United Kingdom played but a minor role, with exports to West Germany, Japan, and other nations accounting for only a few hundred thousand dollars a year. However, as U.S. purchases fell off, those by the United Kingdom became relatively more important.

The effect of the drop in U.S. purchases on the Canadian industry would have been even more severe if it had not been for the stretch-out of U.K. purchases (nearly 1,000 MTU annually through 1971) and intervention by the Canadian government through two large stockpile programs that permitted the principal producers to continue operations. Between 1963 and 1970, Canadian government stockpile purchases totalled 7,400 MTU at a cost of C$101.4 million.[35] This material was held as a general

Figure 5–3.   Value of Canadian Uranium Exports.

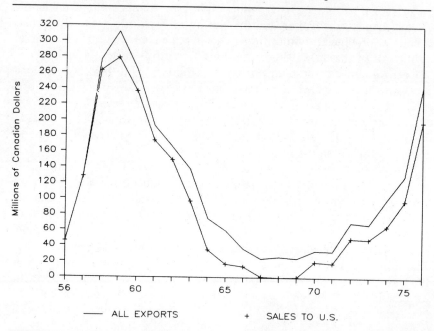

Source: Energy, Mines, and Resources Canada. 1966 (pp. 485–498), 1971 (pp. 433–446), and 1976 (pp. 583–596). *Canadian Minerals Yearbook.* "Uranium" Chapter. Ottawa, Canada.

Note: The unidentified uranium revenues shown were associated primarily with sales to the United Kingdom. Only small shipments were made to Japan, West Germany and other nations, largely for experimental purposes. Revenues prior to 1956 and after 1976 have not been made public.

government stockpile. In order to save Denison Mines, and the community at Elliot Lake, the Canadian government subsequently entered a joint venture agreement that would sustain Denison's production above the level necessary to meet remaining delivery requirements.[36] The program resulted in an additional stockpile of 2,500 MTU over the period 1971–74 at a cost to the government of C$29.5 million. However, this stock was not simply added to the general government stockpile, but rather was controlled by Denison and a new crown company created for the purpose, Uranium Canada, Ltd. (UCAN).

Figure 5–4 shows Canadian uranium production between 1948 and 1982, uranium purchases reported by the U.S. Atomic Energy Commission and the United Kingdom between 1956 and the end of procurement in 1967, as well as early commercial sales (to be discussed below). U.S. purchases

Figure 5-4. Canadian Sales In Transition: Military Procurements, Maintenance Programs, and Commercial Sales.

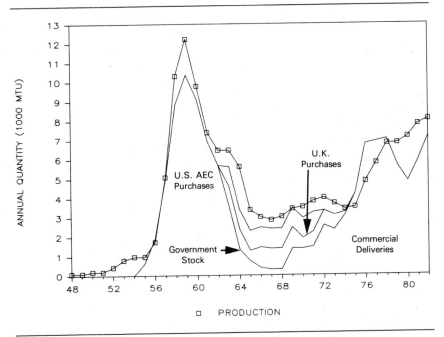

Source: Uranium production from Appendix B; U.S. purchases from: Statistical Data of the Uranium Industry. United States Department of Energy. GJO-100(83), p. 74. U.K. and Canadian government stockpile purchases are author's estimates.

prior to 1956 are not well documented, due to secrecy laws, though there are reports of uranium imports from Canada as early as 1937 and indications of substantial imports prior to 1956.[37] Through 1957, the amount of material unaccounted for in Figure 5-4 totals about 6,700 MTU; most of this material must have been sold to the United States. Exports to the United Kingdom began only in 1958, and it is evident from Figure 5-4 that such purchases were considerably smaller than those of the United States.[38] Deliveries between 1958 and 1962 were reported as being worth C$115 million; at prices then current, this would involve about 5,000 MTU. Between 1963 and 1974, contributions to the three domestic stockpile programs and additional sales to the United Kingdom absorbed more than 19,000 MTU, in addition to the remaining U.S. purchases and early commercial sales. When known commercial sales and domestic reactor requirements (about 1,700 MTU prior to 1975)[39] are taken into account, virtually all of Canada's uranium production is accounted for.

In the following sections, we shall return to this history and to the evolution of Canadian government uranium policy. First, however, we review the structure and activity of the commercial uranium industry in Canada.

Industry, Structure and Contracting

In Canada, uranium production today occurs principally in Ontario and northern Saskatchewan. Uranium deposits in Ontario are relatively low in grade (with uranium averaging less than 0.1% of ore),[40] while those in Saskatchewan are relatively high (fractions of 1% up to about 45%). Early production in Saskatchewan came from the Beaverlodge facilities of Eldorado Nuclear. In Ontario, Rio Algom and Denison Mines still account for more than half of Canada's annual uranium output, from deposits at Elliot Lake.

The number of major producers in Canada has increased rapidly, but the important producers of the past (Denison and Rio Algom) have also increased output. In Ontario, Denison Mines produced 2,360 MTU in 1982, and Rio Algom (expanding and reviving its Quirke and Panel mines) is increasing production from 2,920 MTU in 1981 to 4,460 MTU in 1984. Madawaska Mines, Ltd. revived the Faraday mine in the Bancroft area to produce about 400 MTU annually under a purchase agreement with Italy's Agip, which provided much of the financing. However, operations were shut down in 1982 when Agip stopped its purchases. The largest producer in Ontario—directly or indirectly—is Rio Algom, an affiliate of the Rio Tinto Zinc Corporation, Ltd. (RTZ) of the United Kingdom. Through organizational consolidation of Rio Algom and Preston Mines, RTZ will produce 28,000 MTU over the next forty years for Ontario Hydro and significant additional quantities in connection with major export commitments.

In Saskatchewan, Eldorado has closed its facility near Uranium City, due to high costs and low prices. Rabbit Lake, exploited by a joint venture of Gulf Minerals Canada, Ltd. (a subsidiary of Gulf Oil Corporation of the United States) and Uranerz Canada, Ltd. (a subsidiary of Uranerzbergau, Gmbh., a company with close ties to the West German utility, RWE), attained a production capacity of 1,500 MTU in 1980, although a change in ore bodies caused production to fall to 1,150 MTU in 1981. At Cluff Lake, Amok (a consortium of four French organizations) reached a production level of 1,500 MTU in 1982. And at Key Lake, a mining consortium that includes both provincial and federal agencies—the Saskatchewan Mining Development Corporation (SMDC) and Eldorado Nuclear—is proceeding with a rich new deposit. The locations of these activities are shown in Figure 5–5.

Figure 5-5.   Uranium in North America.

Source: OECD Nuclear Energy Agency and the International Atomic Energy Agency. 1982. *Uranium Resources, Production and Demand*. Paris, France: Organization for Economic Cooperation and Development, p. 209.

In what follows, we present a mine-by-mine review of industry developments in Canada, followed by a discussion of known export commitments and trade patterns.

Agnew Lake Mines.   This wholly owned subsidiary of Kerr Addison has experienced a recurrently unhappy existence. Plans for mining the estimated 9,200 MTU began in the late 1960s, with development work suspended in late 1970 due to the poor uranium market and the high costs that resulted from a low ore grade of 0.05 percent uranium oxide. The revival in the market brought a renewed interest in the mine and rehabilitation began in 1974–1975, with mining beginning in 1976 and the surface plant starting up in June 1977. The operation was originally designed to produce 400 MTU a year, but a host of technical problems raised costs and often reduced production to about half of original plans. The company was unable to solve its technical problems satisfactorily, and declining prices led to a decision in 1979 to stop production. Mining ended in 1980, and although recovery of uranium from mine water and ore stockpiles on the surface continued in 1982, the operation was shut down in 1983. Decommissioning is scheduled for 1984, and the AECB will allow the company to turn over responsibility for the mine site to the government five years later. It has not been determined whether federal or provincial authorities will assume responsibility after October 31, 1989. Agnew Lake Mines held several small contracts, including 175 MTU in commitments for 1980 and 1981. It had borrowed 770 MTU in 1976 from the general government stockpile held by Eldorado Nuclear to meet delivery commitments.[41]

Denison Mines. One of the oldest of the Canadian uranium operations, Denison is currently undergoing a major expansion program, including the rehabilitation of two old mines, as a result of large contracts with Japanese and Spanish utilities, and a thirty-two-year, 48,500 MTU contract with Ontario Hydro, Canada's major utility, which is providing financing for the expansion. High costs, low prices, and other setbacks have not yet materially altered these plans.

Although it still holds 97,000 MTU in reserves, Denison, as one of the mature mining operations in the Elliot Lake region, has suffered from a number of problems, including declining ore grades.[42] This, together with new contract obligations, necessitated a major expansion program to increase milling capacity and ore feed. The expansion of milling capacity from 6,440 tonnes of ore per day (tpd) to 13,610 was set back by a fire in September 1980, which destroyed the new multi-million dollar facility. Despite the delay that resulted, the expansion program was 87 percent complete by the end of 1982. Production increased by 29 percent to 2,360 MTU in 1982 as a result of the expansion program.[43]

The old Stanrock and Canmet mines are being rehabilitated and integrated with Denison. This program, expected to be completed by 1985, was being paid for by an advance from Ontario Hydro of $151 million against future uranium deliveries. However, with the decline in uranium prices, Ontario Hydro has reexamined the economics of both mines, especially as they compare with the high-grade producers in Saskatchewan. Under Ontario Hydro's contract, the utility will pay a cost-based price that will be very high, as much as twice recent spot-market prices.[44] Although the contract contains a cancellation clause that could be invoked after 1985 if the base price exceeds market prices for five years, the penalties would be very high. As a result, Ontario Hydro resorted to legal actions to block development of the Stanrock and Canmet mines. The utility eventually settled out of court, agreeing to allow Denison to develop the mines in return for a 9 percent reduction in the utility's contract purchases (about 4,200 MTU) and a stretch-out in deliveries of one year.[45] As a result, the reopening of Stanrock was delayed until after 1985, when it will be phased in at only a third of originally planned capacity.

Denison's immediate prospects are enhanced by the settlement of the dispute with Ontario Hydro, since that contract covers two-thirds of Denison's future deliveries. However, the possibility of further reductions, especially for deliveries in the 1990s, cannot be excluded. Denison's remaining commitments are primarily to Tokyo Electric Power Co. (Tepco), including a ten-year, 12,880 MTU contract that is near completion, and a ten-year, 15,400 MTU contract that begins in 1984. Some deliveries to Enusa, the Spanish utility, remain, but are small compared to commitments to Ontario Hydro and Tepco. Although the reduction in the Ontario Hydro contract would allow for some new sales, the high costs of production at Elliot Lake would make it difficult to sell uranium at current market prices, especially when low-cost producers like Key Lake can easily undersell Denison.

Rio Algom.   Rio Algom controls three uranium mines at Elliot Lake. Over the past twenty-five years, this company has produced 42,000 MTU and has continuing export and domestic commitments into the next century. In many respects it resembles Denison, since it has high costs and declining ore grades (on the order of 0.1%). It also has an expansion program involving the rehabilitation of two older mines, as well as a contract with Ontario Hydro over which there have been disputes about high prices.

The Quirke mine, where mining began in 1968, was expanded to a production rate of 8,900 tonnes of ore per day by 1978, and the mill capacity expanded to 6,350 tpd. In addition, the Panel mine, which closed in 1961 after only three years of production, was re-opened in 1979 with a production rate of 4,200 tonnes of ore per day, and a mill capacity of 3,000 tpd.[46] The final phase of the expansion began in January 1978, with the signing by Preston Mines Ltd. of a 27,700 MTU contract with Ontario Hydro that included a sizable advance of C$188 million for the rehabilitation of the Stanleigh mine.[47] In January 1980, Preston amalgamated with Rio Algom (under the latter's name).

Prior to this date, Preston and Rio Algom—both affiliates of the United Kingdom's Rio Tinto Zinc—had been related through interlocking directorates. Following the amalgamation, 52.8 percent of the new company was owned by RTZ. The Stanleigh mine was expected to enter into production in July 1983, reaching full capacity in 1984 at 6,350 tonnes of ore per day, with a mill capacity of 4,540 tpd.[46] The company has also conducted studies of the feasibility of re-opening a number of other shut-in mines (Milliken, Lacnor, and Nordic). However, the state of the market makes their rehabilitation unattractive for many years yet.[46,48]

Rio Algom's expansion was required to meet pre-1974 contracts, given declining ore grades. But new contracts were also written. Like Denison Mines, Rio Algom has also had difficulties with its existing contracts. The Tennesse Valley Authority abrogated its 1974 contract for 6,540 MTU. Rio Algom subsequently resold most of this amount.[41]

Rio Aglom has also seen a deferral of deliveries in both its Duke Power and Ontario Hydro contracts, resulting in a reduction in deliveries of several hundred tonnes per year. Duke, which was to receive 7,690 MTU from 1981 to 1990, has delayed part of its scheduled 1981 to 1985 deliveries until 1991 and 1992. Ontario Hydro has exercised its option to defer 15 percent of its deliveries in the first part of the contract period. A 15 percent reduction in production from the Stanleigh mine is planned as a result.

Besides these contracts, Rio Algom has several smaller contracts, including a 1,300 MTU contract with Pruessenelektra, a German utility, a British Nuclear Fuels Ltd. (BNFL) contract for 7,690 MTU, scheduled deliveries of 615 MTU to the Japan Atomic Power Company, and a 1,600 MTU sale to Korea Electric.

Madawaska Mines.   The Faraday mine, which had been taken out of production in 1964, was rehabilitated when Madawaska Mines Ltd. was formed by Federal Resources Corp. (51%) and Consolidated Canadian Faraday (49%) in 1975. A subsequent reorganization led to full ownership of Madawaska Mines Ltd. by Federal Resources, but 49 percent of the ownership of the Faraday mine and facilities was transferred from Madawaska to Consolidated Canadian Faraday.[49] The mine came back into production in 1976.

The Faraday mine was re-opened with a design capacity of 320 MTU per year. However, there was difficulty in delivering enough ore of adequate grade to the mill and, as a result, production had only reached 230 MTU by 1979.[50] Subsequent additions to the workforce and an increase in the number of workplaces in the mine resulted in steady growth in production levels.

Madawaska re-opened on the strength of its contract with Agip, which called for the Italian utility to take 2,300 MTU at a price 5 percent less than the world market price.[51] Deliveries had reached 230 MTU per year under this contract, but Agip, as part of the retrenchment of its nuclear program, renounced its intention to take any more uranium after June 30, 1982. The result was the closing of the Faraday mine in July 1982.

Beaverlodge (Eldorado Nuclear).   Beginning in 1954, the Beaverlodge mine and mill, near Uranium City in northwestern Saskatchewan, eventually produced a total of 17,000 MTU for the crown corporation Eldorado Nuclear.[52,53] Like the Ontario mines, the ore grade at Beaverlodge dropped 40 percent between 1975 and 1980, reaching 0.18 percent. The high costs of operating at depths of a mile made Beaverlodge one of the highest-cost operations in the world, estimated at C$55 per pound in 1980. The company was also hampered by contracts signed in the early 1970s at low prices; these contracts did not expire until 1980.[54]

Despite efforts to increase production by utilizing the adjacent Dubyna deposit in 1979, Eldorado shut down operations in the Beaverlodge area in June 1982. The Atomic Energy Control Board has given approval of the decommissioning plan, the first close-out of a uranium mine under new regulations. Provincial authorities have agreed to monitor the site after the mine has been decommissioned.

Eldorado should have no difficulty in meeting its remaining contract commitments. In addition to its equity share in Key Lake's riches, Eldorado took over the government stockpile of 5,570 MTU in 1981.[55] This takeover was apparently originally intended to increase Eldorado's equity so that it could more easily borrow money to expand its uranium hexafluoride conversion facility. However, Eldorado subsequently transferred almost 4,000 MTU of the stockpile to Gulf (U.S.A.) and Uranerz in return for Gulf's equity in Gulf Minerals Canada, Ltd. (and its holdings at Rabbit Lake and Collins Bay) and Uranerz's share of Rabbit Lake. The state company was thus transformed from an old producer with high-cost production into a competitive low-cost producer, but only with a major government subsidy. Eldorado's customers have included German and Japanese utilities, as well as Spain's Enusa.

Key Lake Mining Corporation. This consortium was originally a joint venture held by the Saskatchewan Mining Development Corporation (SMDC), Uranerz Ltd., and Inexco Ltd., in equal shares. Inexco then sold its share to SMDC, which resold half of it to Eldor Resources, a subsidiary of Eldorado Nuclear. The result was that federal and provincial government organizations ended up with two-thirds of Key Lake's equity. Despite such a high level of government participation (or perhaps because of it), the project has suffered a number of regulatory delays. Lower-grade overburden had been removed by the end of 1982. This discarded overburden contains uranium at an ore grade (0.5%) higher than that for most mines worldwide. Uranium production began in October 1983.

The three deposits at Key Lake are: Deilmann, with estimated reserves of 44,700 MTU at an average ore grade of 2.49 percent; Gaertner, which has 23,000 MTU at a grade of 3.30 percent; and the Boulders, with 2,730 MTU at 0.54 percent.[56] Obviously, these grades tend to make production costs low, though infrastructural factors have led to large initial investments. The original plan called for an annual production capacity of 4,600 MTU to be operational by mid-1983, with the Gaertner ore body being mined first. Full capacity utilization has been delayed until late 1984 by technical problems.

Regulatory questions raised in the Key Lake Board of Inquiry, formed by the provincial government in December 1979, significantly delayed development at Key Lake. Social concerns in the inquiry related primarily to the impact of future mining operations on the local inhabitants, and the desire that they share in the economic benefits. To this end, the board recommended the establishment of:

- an "affirmative action" program to make maximum use of natives, women, and handicapped persons, and of an oversight committee to assure compliance;
- a 50 percent target for native jobs during production;
- a federal board or royal commission to investigate native land claims;
- a technical training program based in the north;
- a compensation program for damage done by developers to traplines; and
- a program of financial aid to northern businesses to help them realize growth opportunities.[56]

These recommendations were intended to ensure that half of the money spent on development would go to the people and businesses in the area. The board expected that taxes and SMDC's share of the profits would account for C$3.3 to C$3.9 billion of the C$6 to C$8 billion total expected lifetime revenue.[56] However, this calculation was based on optimistic uranium prices. After a number of delays that saw SMDC in conflict with other government agencies, a surface lease was signed in August 1981.

Environmental concerns derive in part from the glacial origins of the deposit. Gaertner and Dielman are partially covered by glacial lakes, and the water table is close to the surface. Development of the mine will result in an increase of the flow of the Wheeler River. The company has also found it necessary to provide shielding on equipment to protect workers from radiation from the high-grade ore.

The future production level at Key Lake was still uncertain as of mid-1983. Because the mine and mill will be able to produce at a wide range of levels with no significant loss of efficiency, the company may vary production from year to year to match sales. Expectations of annual production levels range from 3,000 to 4,600 MTU.[57] The disposal of the majority of Key Lake's uranium is still undetermined since sales confirmed by early 1983 amounted to only about 15,400 MTU. Buyers include Ontario Hydro, four U.S. utilities, two Swedish groups, and Belgium's Synatom. Tentative agreements and those not yet approved may total an additional 11,500 MTU to U.S., Japanese, French, and German customers. Key Lake appears to be dominating the marketing or uranium from primary producers, making a number of new sales for amounts ranging from 80 to 380 MTU per year. These sales have been possible in part because of the willingness to write contracts with flexible terms and prices close to spot-market levels.

SMDC has signed contracts with the price based on a sliding percentage of spot-market prices, where the higher the spot-market price, the lower the percentage of that price (down to 92%) the buyer would pay. There was also an absolute ceiling of $35 per pound, adjusted for inflation. Delivery amounts can be adjusted up or down by 15 percent on a year's notice, and a "force majeure" clause allows for contract termination in the event of reactor shutdown or similar problems.[58]

Such attractive contract terms were achieved only after conflicts between federal and provincial interests. The federal government, through its Atomic Energy Control Board, wanted a floor price of about $30 per pound. In contrast, the provincial officials were interested in ensuring sales at competitive prices. With low costs at Key Lake, profitability could be maintained even at the low spot prices of 1982. When Uranerz's 1981 contract with Synatom was delayed in the federal approval process, there was a fear that the Atomic Energy Control Board would disallow the pricing formula or use the annual review of export prices to enforce a floor price higher than that called for in the contract. However, the AECB was ultimately able to accept the complex pricing formula tied to spot-market prices. It also explicitly agreed that prior approval of a pricing mechanism would be taken into consideration during subsequent annual price reviews.

In most Key Lake sales the requirement that uranium exported from Canada be upgraded—through conversion to uranium hexafluoride—was eased. Where many Canadian customers already had "take-or-pay" conversion clauses, Key Lake buyers were allowed to receive the material as yellowcake. Now that the "base load" of Key Lake's production has been sold, government officials have indicated that the requirement will, once again, be enforced. This reportedly caused two U.S. buyers to decide against Canadian uranium in 1983.[59]

Rabbit Lake.   Until mid-1982, this development was jointly owned by Gulf Minerals (45.9%), Gulf Canada (5.1%), and Uranerz Canada Ltd. (49.0%), and was the first of the new Saskatchewan mines to reach production. Original recoverable reserves of ore were estimated at 16,000 MTU, and, at a production rate of 1,700 MTU per year, a ten-year life span was envisioned. Following startup in late 1975, production generally exceeded capacity such that recent output has declined due to a need to shift to new deposits.

Ore grades at Rabbit Lake average 0.4 percent, but some ore has a much higher concentration and requires blending to avoid exceeding the design limit of 0.6 percent at the mill. Despite the cost of blending, of draining half of Rabbit Lake, and the need to reduce worker turnover by providing frequent leave in more hospitable areas, production costs at Rabbit Lake are among the lowest in the world.[60] Since Uranerz agreed to buy up to 770

MTU annually above its equity share, Gulf was virtually guaranteed a good return. Gulf and Uranerz sold most of the mine's output to U.S., Japanese, and European utilities.[61]

With depletion approaching at Rabbit Lake, both Gulf and Uranerz were active in exploration. Gulf discovered the Collins Bay "B" deposit containing over 19,000 MTU in recoverable reserves, and the Eagle Point deposit holding more than 7,000 MTU. Original plans called for mining Collins Bay as early as 1982, but despite environmental approval in 1982, exploitation has been delayed until 1985. As a result, production at Rabbit Lake dropped to 1,150 MTU annually from 1981 to 1984. Although the presence of metal oxides and arsenides in the Collins Bay "B" ore will necessitate modfications of the mill at Rabbit Lake, production costs are expected to remain very low.[62]

Eldorado Nuclear, the federal uranium corporation, and Gulf Oil Corporation (100% owner of Gulf Minerals Canada Ltd., and 60% owner of Gulf Canada) agreed late in 1982 on a transfer of Gulf's uranium holdings—including the Collins Bay and Eagle Point deposits as well as its share of the mill at Rabbit Lake—in exchange for uranium from Eldorado's 5,400 MTU stockpile. Less than 400 MTU was given to Uranerz for its share of the mill, and about 3,600 MTU went to Gulf.[63] Gulf will probably use some of the uranium to meet deliveries required under the settlement of the Westinghouse lawsuit, following shutdown of Gulf's Mt. Taylor Mine.

Saskatchewan's Mineral Resources Minister in 1982, Colin Thatcher, was opposed to the buyout by Eldorado. He argued that the massive infusion of uranium into the United States would bolster arguments in favor of an American uranium import embargo or quota, much to the detriment of Canadian producers. He also suggested that it would be best to retain significant private involvement in the uranium industry in the province to provide a balance to government-owned or government-controlled operations.[64] However, it is also likely that Saskatchewan officials wanted to minimize federal control over the provincial industry.

Despite Thatcher's opposition, the federal government approved the deal in September 1982, and the uranium was subsequently transferred to Gulf in the United States. A new company, Eldor Mines Ltd., will take over Gulf and Uranerz operations in the area around Rabbit Lake, as well as their contractual obligations (to Hokuriku Electric of Japan and Ontario Hydro). The AECB has granted approval for a 1,920 MTU annual production level.

Cluff Lake.   This project, originally owned entirely by a consortium of French companies, Amok Ltd. (Mokta at 25%, Pechiney Ugine Kuhlmann at 25%, CMFU at 20%, and CEA at 30%), now includes SMDC as a partner, with a 20 percent holding.[65,66] Four deposits are included at Cluff Lake: "D," which contains over 5,000 MTU in reserves and averages 7 percent in ore grade; "N," which has 5,000 MTU in reserves; Claude, with 4,800 MTU of reserves; and the "O-P" deposit, which has 1,500 MTU in reserves. The latter two deposits have concentrations of about 0.4–0.7 percent uranium oxide, while the "N" deposit averages 0.3 percent.[67,68] The ore grade in the "D" deposit approaches 45 percent at times, and requires special shielding for the operators.

From June 1980 through October 1981, the "D" ore body was mined, creating a surface stockpile for Phase I production. Milling of the ore began in October 1980 and quickly reached an output rate of 1,500 MTU per year. Milling will continue until mid-1984, followed by the recovery of 500 to 700 tonnes of uranium from gravimetric residues. A feasibility study is being conducted concerning a Phase II operation involving development of the "O-P," Claude, and "N" ore bodies, as well as a major mill expansion.[69] The latter would be necessary because of the significantly lower ore grades in the remaining deposits. Phase II would allow production, at a lower level into the early 1990s, with open-pit mining at the Claude deposit and underground mining at the "O-P" deposit. The French state enterprise Cogema has agreed to take 80 percent of the projected 960 MTU Phase II annual production.

The uranium produced under Phase I at Cluff Lake was equally divided between Cogema and an unnamed German utility, with Amok and SMDC splitting the revenue in proportion to their equity shares. Given 1982 production of 1,900 STU_3O_8 and SMDC sales of C$24.7 million, a price of C$32.5 per pound can be derived, somewhat less than the C$35.50 per pound average export price reported by the AECB for 1982.[69,70]

Canadian Export Commitments

Canada's cumulative and prospective exports of uranium are the largest of any producer. Since commercial contracting began in 1966, Canadian producers have entered into arrangements to export about 132,000 MTU.[71] Of this, about 61,000 MTU were exported prior to 1983, leaving a forward commitment of at least 71,000 MTU.

As indicated in Figure 5–6, Japan is Canada's largest customer, receiving 36.5 percent of lifetime export commitments; the United States is next with 17.1 percent, followed closely by the United Kingdom (11.6%) and West Germany (12.0%), Spain (6.1%), and France (6.6%). Sweden, Belgium, South Korea, Finland, Italy, and Switzerland (in descending order) receive the remaining 10 percent. Current indications are that the U.S. share will increase significantly from 1983 on.

Under Canadian law, export commitments made since September 1974 must be approved by the Atomic Energy Control Board for conformity to

Figure 5-6.   Canada's Commercial Uranium Exports (*Estimated*).

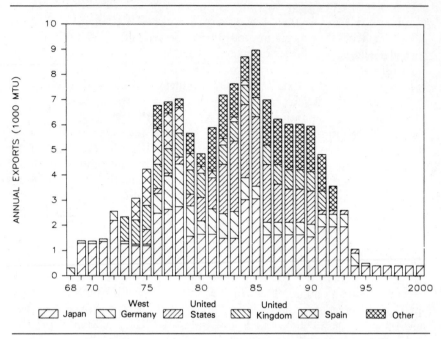

Note: Does not include the 1982 transfer of Canadian government stocks to Gulf Corporation in the United States as part of a trade for Gulf properties in Canada.

nonproliferation conditions, price, arrangements for value added in Canada (primarily UF_6 conversion), and other conditions (such as the requirement that sufficient reserves be held for domestic needs).[72] Approvals granted as of the end of 1982 totaled about 86,000 MTU. Thus about 51,000 MTU either (1) was committed prior to September 1974, (2) extends beyond the official ten-year approval horizon, or (3) has not yet received approval on export conditions. Except for some recently negotiated deals, there appear to be few major contracts in the third category. However, the AECB has delayed contract fulfillment in at least a few instances, until price or conversion requirements were met. An example is the Madawaska contract with Agip (Italy). Agip had provided major mine financing and expected to receive uranium at about $32 per pound; Canadian officials insisted on an increase to about $42 per pound. More recently, Belgium's Synatom experienced delays in obtaining export approval of its contract for Key Lake uranium during government review of its complex pricing formula tied to spot-market prices.

Policy Context

Canadian uranium export policy has been shaped over a period of nearly thirty years by a series of domestic and international events. Internationally, these events have focused primarily on efforts to secure markets for Canadian exports and on safeguards and other nonproliferation issues. Domestically, the issues have centered on control of natural resources and allocation of the wealth deriving from such resources. The latter have arisen in the context of long-standing conflicts between the provinces and the federal government over resources, concerns about foreign ownership and control of Canadian resources, and domestic energy security. The evolution of Canadian uranium policy reflects all of these influences, and while an examination of the historical development under such conditions can be confusing, it is perhaps the only way to weigh various contributing factors. In the following, we review this history by emphasizing those factors most important at each stage, beginning with events at the national and international levels.

As discussed earlier, federal authorities intervened in the years between military procurements and commercial sales to support the domestic uranium industry through stockpile purchases and joint ventures. However, other federal actions inhibited commercial activity. For example, Denison Mines was negotiating a sale with France when, in June of 1965, the government of Canada implemented a new policy restricting exports to material to be used only for peaceful purposes. Anticipating actions to be taken much later, the statement asserted that

> Before such sales to any destination are authorized, the Government will require an agreement with the government of the importing country to ensure with appropriate verification and control, that the uranium is to be used for peaceful purposes only.[73]

This condition was not acceptable to France, and exports of Canadian uranium to that nation did not occur until 1981.

In June 1969, new restrictions were placed on Canadian exports. As announced by the Minister of Energy, Mines and Resources, the new policy required that

> . . . all contracts covering the export of uranium or thorium be examined and approved by the appropriate federal agency before any application for an export permit is considered. The examination will cover all aspects and implications of the contract such as nuclear safeguards, the relationship between contracting parties, reserves, rate of exploitation, domestic requirements, domestic processing facilities, and selling and pricing policy. Approval will not normally be given to contracts of more than ten years duration unless provision is made for renegotiation of price.

Once an export contract is approved, export permits covering the actual shipment of uranium or thorium may be issued annually provided that the conditions of the contract have been maintained. Approval will be granted only for the supply of uranium and thorium for peaceful purposes to customers in countries with which Canada has completed a safeguards agreement, or following the coming into effect of the Non-Proliferation Treaty, with customers in countries which have concluded the necessary safeguards agreement with the International Atomic Energy Agency.[74]

These regulations dealing with trade issues were supplemented by notices concerning ownership of Canadian resources, following proposals for the sale of a controlling interest in Denison Mines to Hudson's Bay Oil and Gas, a company with foreign ownership. In March 1970, the Canadian government indicated that it would seek legislation to limit foreign ownership of mining operations (exploration would not be so constrained) to 33 percent. Foreign holders of greater shares of existing operations as of March 2, 1970, would be able to sell such shares only to Canadian residents until foreign ownership was reduced to the 33 percent limit.[75]

While this regulatory framework was being developed, Denison and other Canadian producers were seeking export sales opportunities. Commercial sales to U.S. utilities—the largest market at the time—had been foreclosed by U.S. law in 1964 (effective in 1966) that prohibited enrichment of foreign uranium for use in domestic reactors.

Opportunities to sell to non-U.S. utilities were also reduced by the 1971 proposal (implemented in 1972 for deliveries beginning in 1973) of the U.S. AEC to reduce its stockpile by operating enrichment plants at 0.275 to 0.30 percent tails assay, while requiring customers to deliver the smaller amounts of uranium that would be needed at a "transaction" tails assay of 0.20 percent. Based on U.S. exports of more than 55,000 MTSWU over the decade beginning in 1973, the split-tails policy would essentially displace as much as 25,000 MTU of non-U.S. uranium demand during that period, compared to what would be required if the United States required uranium deliveries commensurate with a higher tails assay.

Canada's first commercial export sales were made in 1972 when Denison and UCAN arranged to sell about 3,600 MTU to a group of Spanish utilities. This was more than 1,100 MTU greater than the joint stockpile and was to be made up from government stocks. Denison also agreed to sell 770 MTU to Tohoku Electric of Japan, also to come from Canadian government stocks. These and subsequent sales, however, took place during a period in which Canadian government and industry officials were meeting with counterparts in other producing nations to arrange international terms of trade in uranium.

The Cartel.   Meetings of what later was called a cartel took place in Paris in February and April 1972, at the end of May in Johannesburg, in Cannes in July, and at other locations subsequently. As a result of these meetings, Canada was allocated 33.5 percent of the world market (excluding the home markets of the United States, France, Canada, South Africa, and Australia) for the period 1972-77, and 23.22 percent for 1978-80. The contracts with Spain and Japan were considered part of Canada's quota, but a fraction of Gulf-Uranerz production at Rabbit Lake was excluded on the basis that it was captive production destined for West Germany.[76]

The cartel also set uranium prices, imposed bidding rules, and other necessary details for allocating demand. The Canadian government was deeply involved in cartel activities and even notified Washington and other foreign government offices of the February meeting. It also extended the legal basis for exports described above to accommodate price and allocation rules established by the cartel. A new regulation under the Atomic Energy Control Act was added in July 1972 which reads as follows:

> A permit to export prescribed substances shall not be granted unless the [Atomic Energy Control] Board is satisfied that the prices stipulated for, and the quantities of, the prescribed substance proposed to be exported meet such criteria, if any, respecting price levels and quantities as may be specified in the public interest in a direction given to the Board by the Minister.[77]

Subsequent directives to the AECB specified how Canada's quota would be allocated among domestic producers and the schedule of prices.[78,79]

The first directive was issued in August 1972 with prices for 1973 deliveries put at $5.75 per pound, except for sales to Japan, Korea, and Taiwan which were to pay $5.95 (with prices higher due to what were termed "commissions"). Deliveries in 1978 were to be priced at $7.50 ($7.70 for Asian buyers). Prices were revised upward in May, August, and October of 1973 and again in March of 1974 (when prices for 1974 deliveries were set at $8.20 and for 1978 at $12.20). These revisions appear to have resulted not so much from efforts of the cartel to raise prices further as from the need to keep up with rising uranium market prices that consistently outstripped government adjustments. As we have argued, there were many factors that contributed generally to rising prices during this period.

By early 1974, uranium market conditions had fully reversed with prices exceeding $12 per pound, well above the government-specified price. During the first six months of 1974, Canadian producers contracted for forward deliveries of about 35,000 MTU.[80] This was nearly ten times the annual production rate at the time. In January 1974, the Minister of Energy, Mines and Resources voiced concern about the sufficiency of uranium supplies for domestic nuclear power use. The concern was expressed in policy in September 1974 when uranium producers were asked to set aside a portion of reserves (based on expected nuclear growth and the producer's share of national reserves), and domestic utilities were required to establish contracts for at least fifteen years forward supply.[72] Of course, these moves also constricted potential supply from Canada and may have contributed to further world uranium price increases.

At the same time, the terms of the 1969 policy were reiterated, with the additional condition that export approval would also be contingent on producer conformity to domestic reserve requirements. In addition, what had been an informal policy was made a formal requirement in that export of uranium be "made in the most advanced form possible in Canada," unless special exemption were granted. In practice, this has meant conversion to uranium hexafluoride, though there have also been proposals to build enrichment facilities in Canada.

The September announcement also formalized the export-approval process, with producers being "required to submit their sales contracts to the Atomic Energy Control Board for examination prior to being granted a licence to export the uranium." The factors to be examined included "basic pricing, quantity and delivery conditions." Notice was given in the September statement that export approval would not be given "if the pricing conditions for foreign customers are more favorable than those offered to domestic purchasers."[72]

Nonproliferation Imperatives.   However, it was also in 1974 that this growing preoccupation with commercial affairs was shaken by international events. On May 18, India exploded a nuclear device beneath the Rajasthan desert. With the plutonium for this device apparently coming from a natural uranium reactor supplied by Canada, Canadian reaction was inevitable.[81] Four days later, Canada suspended nuclear assistance to India and began a critical re-examination of its nuclear export policies. This re-examination culminated in December 1974 in new safeguards requirements for Canadian customers. Not only were IAEA safeguards to be required on all Canadian-supplied material and technology (and derivatives), but Canada called for renegotiation of existing cooperation agreements, a right of prior approval on retransfers, and retroactive imposition of all conditions on uranium supply contracts already approved.[82]

A one-year period was set for renegotiation, during which time uranium trade would continue normally. When, at the end of 1975, renegotiations had not succeeded, the deadline was extended until mid-1976 and then until the end of 1976. Continuing failure to achieve Canada's conditions resulted at the end of 1976 in an embargo of supply to Japan, the United States, Switzerland, and the Euratom nations, as well as a new demand that Canadian uranium customers ratify the NPT or otherwise provide for safeguards on their entire nuclear programs.[83] These unilaterally imposed conditions were not acceptable to many of Canada's customers. However, as described in Chapter 2, the new U.S. administration soon reassumed the lead in nonproliferation matters, taking the spotlight off of Canada not only by its greater visibility but also through temporary accommodations afforded by the International Nuclear Fuel Cycle Evaluation activity.[84]

With the United States taking the lead on nonproliferation, Canadian attention returned to economic issues associated with exports. In March of 1977, the Minister of Energy, Mines and Resources announced that:

> With regard to all contracts not yet approved by the Atomic Energy Control Board and all future contracts, the government will expect that terms of sale will provide for an annual renegotiation of price based on then existing world prices giving consideration to such factors as term and size of contract and any special financing arrangements. Provision should also be made for an escalating floor price which will protect investment in production facilities.[85]

This step was in response to the fact that exports under some old contracts were increasingly at prices far below what the AECB was willing to approve for new exports. The announcement of this new policy—in close proximity to shifts in nonproliferation policy—caused considerable animosity among Canadian customers, with some linking Canadian nonproliferation initiatives to economic motives.

These conditions on the marketing of Canadian uranium, incrementally developed over a period of some years, appear to put government in control of many aspects of uranium trade, including price, nonproliferation conditions, domestic upgrading, foreign equity participation, and others. However, the exercise of these powers has revealed a lighter hand and a greater pragmatism than might have been expected. And with time, the extent of federal power has been eroded by weakening world nuclear market conditions. At many turns in the preceding developments at the national level, provincial interests have been asserted, increasingly in ways that affect national policy implementation.

Natural Resources and Federal-Provincial Relations.   Under the British North America Act of 1867, natural resources were established as being under both the ownership and control of the original provinces (this power was extended to Saskatchewan in 1930 when the 1867 act was amended). However, provisions in the act allow the federal government to claim jurisdiction over "works and undertakings connecting the province with any other or others of the provinces, or extending beyond the limits of the province" (Section 92 [10a]) and "such works as, although wholly situate within the Province, are before or after their execution declared by the Parliament of Canada to be for the general advantage of Canada or for the advantage of two or more of the provinces" (Section 92 [10c]).[86]

It was on such a parliamentary declaration of "general advantage" that Article 17 of the 1946 Atomic Energy Act claimed control over:

All works and undertakings whether heretofore constructed or hereafter to be constructed,

(a)   for the production, use and application of atomic energy
(b)   for research or investigation with respect to atomic energy, and
(c)   for the production, refining or treatment of prescribed substances,

are and each of them is declared to be works or a work for the general advantage of Canada.[87]

However, since the passage of the Atomic Energy Act there have been successive reinterpretations and elaborations of federal powers and recurrent contests with provincial authorities.

While the provinces have at times been willing to accede to national concerns in regard to strategic issues, they have also sought to limit such federal encroachment. Under the Atomic Energy Act, the provinces still *own* uranium resources and have considerable rights to regulate, tax, and otherwise participate in uranium development. These powers have been used to establish a significant provincial presence in the uranium industry, particularly in Saskatchewan as described above. The provinces have also frustrated several legislative efforts by Ottawa to encode regulations and other elaborations of the federal policy approach begun in the Atomic Energy Act.[88]

The provinces have also been willing to engage the national government even on international strategic issues such as proliferation. For example, Saskatchewan's Cluff Lake Board of Inquiry argued in 1977 that nuclear proliferation can only be solved by a comprehensive, multinational effort and that unilateral actions are ineffective if not counterproductive. The board warned that the withholding of uranium from world markets would probably have harmful effects for nonproliferation; also such a move would conflict with attempts to ameliorate global energy problems. According to the board, it was incomprehensible to speak of a Canadian contribution to proliferation through its uranium exports: "Proliferation exists because of the security structure, not because of Canadian uranium."[89]

The board went further in criticizing the philosophical foundations of the nonproliferation policies articulated by the United States and other countries, including—by implication—the Canadian federal government. A few excerpts from the report make this clear:

- Both sides feared that nuclear weapons could get into the hands of less responsible governments but the supporters of nuclear power claimed that they would refrain from making nuclear materials available to any nation likely to experience a civil war or subnational coup while the opponents of nuclear power wanted Third World nations to develop alternative sources of energy. Nations in the Third World know that both superpowers had serious civil wars themselves and that unstable or dictatorial governments are by no means the special preserve of underdeveloped and developing nations. We agreed that assumptions of inferior qualifications for nuclear energy did sound like a modern version of the White Man's Burden.

- Both sides of the nuclear debate accepted in their evidence that the balance of nuclear terror between the superpowers had helped to keep the world free of a major war for over three decades. We concluded that the nonaligned nations of the Third World considered that international security for them would be as well preserved by a balance of nuclear terror in which they had a reasonable share.

- Both protagonists before us argued for international justice. The pronuclear witnesses argued for the right to make nuclear energy available to the Third World for peaceful purposes under strict safeguards to prevent making nuclear weapons. The non-nuclear witnesses argued that nuclear power was unsuitable for many of those nations and that the traditional non-industrial way of life should be preserved. The Third World considers that these arguments mask the real reason—any proliferation would be a shift in the distribution of power in international decisionmaking. We agreed that it is a redistribution of this power which they want. International justice and equality requires not simply a redistribution of wealth or resources, but also of global prestige, bargaining power in political and economic agreements, and a voice in international organizations. We deplored the fact that nuclear weapons are used as a measure in the allocation of power in the world community.[89]

In making its arguments, the board, and the provincial government that endorsed and implemented the report, show more sympathy for the developing-country point of view than for that of the Canadian federal government. With its heavy dependence on agriculture and natural resources, Saskatchewan shares other common ground with the developing world. However, the province is also part of the Canadian confederation, and perspectives on resource and export policy must ultimately be reconciled.

Canadian interests are united in wishing a maximum return for natural resources; the ability to maintain prices would help fulfill this desire. The difficulties in doing so, however, are both internal and external. Internally, it is difficult to control producers: as world prices drop below levels judged appropriate by the AECB, there are great pressures from low-cost producers in Saskatchewan to go ahead with exports even at lower prices. There is also a tension between desires to yield on price in order to secure a larger share of the highly contested export market and the need to maintain equity between domestic utilities that have high-cost and high-priced contracts with producers in eastern Canada and foreign buyers. This interprovincial equity issue as it relates to energy exports has repeatedly strained the Canadian governmental system.[90-95]

As uranium market conditions and nuclear technology export markets have become less advantageous, Canadian officials appear to have retreated from confronting these basic economic issues. While there is clearly temptation (and historical precedent) to follow the Australian lead in setting a rigorous floor price and being willing to sacrifice some potential sales opportunities in order to secure higher prices, the federal government has acceded to provincial desires to export even at relatively low prices.

However, there are serious and as yet unanswered questions about how national and provincial interests will be balanced in the longer term, given the aggressiveness of provincial authorities and their direct involvement in uranium activities with important local benefits even at today's prices. It should be noted, however, that the annual price review provision will give the national government new opportunities in the future to demand higher prices or to impose other conditions if and when market conditions permit. Thus, even for existing contracts there will be uncertainty, and the recurrent possibility that swings in relative national and provincial power will alter the terms of access to Canadian uranium.

REFERENCES

1. Wright, William. 1981. "Historical Background to Uranium Developments in Australia." Paper presented to the Australian Uranium Symposium (sponsored by Edlow International Co.), Washington, D.C.
2. *Uranium: Australia's Decision.* Background Paper. Canberra, Australia: C.J. Thompson, Commonwealth Government Printer.
3. Statement from the Prime Minister. August 25, 1977. *Australian Foreign Affairs Register,* September 1977.
4. *Nuclear Engineering International.* May 1980. "Mining Delay Costs Sales." p. 4.
5. *Financial Times.* (London) September 1, 1982.
6. Munro, J.G. September 1981. "Overview of Australian Uranium Developments." Paper presented at the International Conference on Uranium, Atomic Industrial Forum/Canadian Nuclear Association, Quebec City, Canada, September 15-18, 1981.
7. *Nuclear Exchange Corporation.* October 1980. "Nuexco Monthly Report to the Nuclear Industry." Menlo Park, CA: NUEXCO.
8. *Nuclear Fuel.* September 15, 1980. Japanese Ranger Partners form JAURD." p. 12.
9. *Register of Australian Mining.* 1981. London: Mining Journal Books.
10. *Nuclear Fuel.* November 24, 1980. "ERA Makes Australia's Top 10 as Ranger Uranium Goes Public." p. 4.
11. *Nuclear Fuel.* October 26, 1981. "Union Stops 280 Tonne Shipment From Ranger As Miners Look For Ways To Break Blockade." p. 5.

12. *Nuclear Fuel.* August 30, 1982. "Ranger's First Year Produces Nice Profit." p. 16.
13. *Nuclear Fuel.* April 25, 1983. "Japanese Utilities Agree to Pay $32–35/Lb. for Uranium in 1983 and 1984." p. 1.
14. *Nuclear Fuel.* November 8, 1982. "Roxby Planners Base EIS on 3,000 tonne/year Uranium Output from $1.4-Billion Mine." pp. 9–10.
15. *Register of Australian Mining.* 1981. London: Mining Journal Books. p. 120.
16. Livingston, R.S. September 1982. "Can Market Stability be Improved? Government and Export Contract Approvals." Uranium Institute: Seventh International Symposium, Session V. London: Uranium Institute.
17. *Nuclear Fuel.* June 20, 1983. "Hawke Bans Uranium Exports to France As Pressure Builds From Labor Factions." p. 3–4.
18. *Mining Journal.* July 3, 1981. "Australia-Northern Wealth," 297, no. 7611: pp. 1–3.
19. *Nuclear Fuel.* June 22, 1981. "Pancontinental Changes Jabiluka Plans." pp. 8–9.
20. *Nuclear Fuel.* May 24, 1982. "Western Mining Sees Market Opportunities in Esso's Withdrawal from Yeelirrie." pp. 7–8.
21. *Nuclear Fuel.* May 11, 1981. "Dispute Between Hydro and Denison Seen Jeopardizing 126-Million-Pound Contract." pp. 1–2.
22. *Oil and Gas Journal.* September 28, 1981. Tulsa, Okla.: Petroleum Publishing Co. p. 152.
23. *Register of Australian Mining.* 1981. London: Mining Journal Books. p. 124.
24. Brooks, James A. 1981. "Australia—A Major Uranium Supplier: Government Policies and Attitudes to Development and Trade." Paper delivered to symposium sponsored by Edlow International Co., Washington, D.C., March 5, 1981.
25. McGregor, W. Graham. September 1981. "The Legislative and Policy Framework for the Development of the Australian Uranium Industry." Paper presented to the International Conference on Uranium, Atomic Industrial Forum/Canadian Nuclear Association, Quebec City, Canada, September 15–18, 1981.
26. Daily Hansard. Parliamentary Debates. Australian House of Representatives. November 27, 1980. pp. 136–143.
27. Department of Treasury. 1981. *Your Investment in Australia, A Guide for Investors.* Commonwealth of Australia.
28. Sharman, Cambell. 1980. "Fraser, the States and Federalism." *The Australian Quarterly.* Sydney, Australia: The Australian Institute of Political Science.
29. Statement of the Hon. J.D. Anthony, Deputy Prime Minister, Minister for Natural Resources and Minister for Overseas Trade. *Uranium: Australia's Decision.* Canberra, Australia: C.J. Thompson, Commonwealth Government Printer.
30. *Asian Wall Street Journal.* December 14, 1981.
31. Neff, Thomas L., and Henry D. Jacoby. 1979. "Supply Assurance in the Nuclear Fuel Cycle." *Annual Review of Energy* 4:259–311.
32. *Nuclear Fuel.* July 19, 1982. "Australia Labor Party Softens Uranium Stand But 'Sloppy Wording' Leaves Many Questions." pp. 5–6.

33. See Figure 3-4 and Appendix B.
34. Energy, Mines, and Resources Canada. 1966 (pp. 485–498), 1971 (pp. 433–446), and 1976 (pp. 583–596). *Canadian Minerals Yearbook.* "Uranium" Chapter. Ottawa, Canada: Energy, Mines, and Resources Canada.
35. Runnalls, O.J.C. 1977. "The Uranium Industry in Canada." Brief submitted to the Cluff Lake Board of Inquiry, Regina, Saskatchewan. Ottawa, Canada: Department of Energy, Mines and Resources, Energy Development Sector. p. 9.
36. Energy, Mines, and Resources Canada. 1971. "Uranium" Chapter, Table 4: "Major Canadian Uranium Commitments Announced Since 1966." In *Canadian Minerals Yearbook.* Ottawa, Canada: Energy, Mines, and Resources Canada. pp. 433–446.
37. U.S. Department of the Interior, Bureau of Mines. 1946 (p. 1,224), and 1947-1956 (various issues). "Uranium" Chapter. In *Minerals Yearbook.* Washington, D.C.: Department of the Interior.
38. U.S. Department of the Interior, Bureau of Mines. 1957. "World Review—Canada" Section. *Minerals Yearbook.* p. 1,234.
39. Computed according to the fuel-cycle requirements given in Chapter 1, and the historic nuclear power growth scenario of Chapter 4.
40. For a description of the geologic characteristics of Ontario deposits and a review of the relevant literature, see: Robertson, James A. 1981. "The Uranium Deposits of Ontario—Their Distribution and Classification." Ontario, Canada: Ontario Ministry of Natural Resources. Ontario Geological Survey Miscellaneous Paper 86.
41. *Canadian Mines Handbook 1980-81.* 1981. Toronto, Ontario: Northern Miner Press Ltd. p. 140.
42. As of the beginning of 1980. *Nuclear Fuel.* August 4, 1980. "Survivors Must Expand to Stay Even As Elliot Lake Ore Grade Declines." p. 9.
43. *Nuclear Fuel.* January 31, 1983. "Denison says Output is Up and Costs Down." pp. 9-10.
44. *Nuclear Fuel.* May 11, 1981. "Dispute Between Hydro and Denison Seen Jeopardizing 126-Million-Pound Contract." pp. 1-2.
45. *Nuclear Fuel.* October 12, 1981. "Settlement With Denison Puts Ontario Hydro in the Market for Another 12-Million Pounds." pp. 1-2.
46. *Nuclear Fuel.* August 4, 1980. "Survivors Must Expand to Stay Even As Elliot Lake Ore Grade Declines." pp. 9-10.
47. *Nuclear Fuel.* May 26, 1980. "Ontario to Review Hydro's U_3O_8 Contracts Because of Payments Now $13/Lb. Over Spot." pp. 12-14.
48. Merlin, H.B. September 1981. "Canada's Uranium Supply Potential." Paper presented to the International Conference on Uranium, Atomic Forum/Canada Nuclear Association, Quebec City, September 15-18, 1981. See pp. 11-12 for discussion of expansion plans for the three mines.
49. Whillans, R.T. 1981. "Uranium." *Canadian Mining Journal.* February: pp. 146-155.
50. *Canadian Mines Handbook 1980-81.* 1981. Toronto, Ontario: Northern Miner Press Ltd. p. 170.
51. *Foreign Uranium Supply.* 1978. Rockville, Md.: NUS Corporation. pp. 3-102.

52. *Canadian Mines Handbook 1980–81*. 1981. Toronto, Ontario: Northern Miner Press Ltd. pp. 98–99.
53. *Nuclear Fuel*. April 14, 1980. "Hard Year for Some Uranium Producers Appears to be Only the First of Many." pp. 10–12.
54. *Mining Journal*. November 30, 1979. "Eldorado Nuclear—Profits Slashed." p. 467; August 29, 1980. "Eldorado Nuclear—New Refinery Location Switched." p. 173.
55. *Nuclear Fuel.* March 9, 1981. "Gift of Stockpile to Eldorado Shakes Industry." Special Issue. pp. 1–2; May 10, 1982. "Gulf's Deal With Eldorado Nuclear Is Seen As Prelude To Exit From Uranium Business." pp. 15–16.
56. *Nuclear Fuel*. February 16, 1981. "Key Lake Developers Will Invest $1-Billion To Get $6- to $8-Billion, Panel Estimates." pp. 10–11. Board estimated prices of $35-$40/lb. U_3O_8 through 1985, and 4 percent annual real increases thereafter.
57. *Nuclear Fuel*. December 20, 1982. "SMDC Calls Key Lake a 'Financial Bonanza' But Finds Cluff Lake Uranium Hard to Move." pp. 15–16.
58. *Nuclear Fuel*. April 11, 1983. "SMDC Provides Up to 8% Off NUEXCO Price and $35/Lb. Ceiling in Yankee U_3O_8Deal." pp. 1–4.
59. *Nuclear Fuel*. July 4, 1983. "Upgrading Policy Hurts Canadian U_3O_8 Sales But Might Change Ways of U.S. Convertors." p. 8.
60. *Nuclear Fuel*. August 31, 1981. "Settlement With Exxon Transforms Gulf From Uranium Buyer to Seller." p. 5.
61. U.S. House of Representatives. December 8, 1977. *The International Uranium Cartel*. Hearing before the Committee on Interstate and Foreign Commerce. Vol. II: pp. 287–292. Serial 95-95.
62. *Canadian Mines Handbook 1980–81*. 1981. Toronto, Ontario: Northern Miner Press Ltd. p. 28.
63. *Nuclear Fuel*. October 11, 1982. "Gulf Receives Most of Eldorado Stockpile Following Canadian Government Approval." pp. 11–12; May 9, 1983. "Eldorado Spared 'Substantial' Loss in '82 by Trading U_3O_8 for GMCL and Uranerz." pp. 15–16.
64. *Nuclear Fuel*. July 19, 1982. "Canadian Government Sees No Point in 'Fighting the Market'—Just Now." pp. 3–4; August 30, 1982. "Gulf Eldorado Directors Agree on Deal and Ask Canadian Government to Approve." pp. 1–2.
65. *Mining Journal*. August 3, 1979. "SMDC Takes 20% Stake in Cluff Lake." p. 90.
66. *Nuclear Fuel*. January 7, 1980. "Public Hearings, Political Confrontation Color Yellowcake Picture in Saskatchewan." pp. 7–8.
67. *Mining Journal*. September 18, 1981. "Phase I Mining Almost Complete at Cluff Lake." pp. 201–203.
68. *Foreign Uranium Supply Update-1980*. Rockville, Md.: NUS Corporation. pp. 3–32.
69. *Nuclear Fuel*. April 25, 1983. "SMDC Writes Off $11-Million in Prospecting." pp. 14–15.
70. Whillans, R.T. 1983. "Uranium." *Canadian Mining Journal*. February: pp. 123–128.

168   THE INTERNATIONAL URANIUM MARKET

71. These are the quantities for which we have been able to find confirmation. There may be some recent spot sales that do not appear in our tabulations or older commitments that have escaped our discovery efforts. In addition, the transfer of as much as 4,000 MTU to Gulf in the United States, as well as a number of as yet unconfirmed sales of Key Lake uranium, have not been included.

72. MacDonald, Donald S. Minister of Energy, Mines and Resources on Canadian Uranium Policy. Statement on Canada's Uranium Policy. September 5, 1974.

73. "House of Commons Debates—Official Report." June 3, 1965. pp. 1948–49.

74. Lang, Otto E. Minister of Energy, Mines and Resources. Statement before House of Commons. June 19, 1969.

75. Greene, J.J. Minister of Energy, Mines and Resources. Statement Regarding Foreign Ownership in the Canadian Uranium Industry. March 19, 1970.

76. U.S. House of Representatives. December 8, 1977. *The International Uranium Cartel.* Hearing before the Committee on Interstate and Foreign Commerce. Vol. II. Serial 95–95.

77. Runnalls, O.J.C. 1981. "Ontario's Uranium Mining Industry—Past, Present, and Future." Ontario, Canada: Ontario Ministry of Natural Resources.

78. These directives are reproduced as Appendices 10–15 of "Ontario's Uranium Mining Industry—Past, Present, and Future," by O.J.C. Runnalls. Ontario Ministry of Natural Resources. pp. 95–101.

79. U.S. House of Representatives. December 8, 1977. *The International Uranium Cartel.* Hearing before the Committee on Interstate and Foreign Commerce. Vol. II. Serial 95–95.

80. Runnalls, O.J.C. 1977. "The Uranium Industry in Canada." Brief submitted to the Cluff Lake Board of Inquiry, Regina, Saskatchewan. Ottawa, Canada: Department of Energy, Mines and Resources, Energy Development Sector. p. 9.

81. For additional discussion of the subsequent events in Canada, see: Morrison, Robert W., and Edward F. Wonder. 1978. "Canada's Nuclear Export Policy." *Carleton International Studies* III. Carleton University, Ottawa, Canada: The Norman Patterson School of International Affairs.

82. MacDonald, Donald S. Minister of Energy, Mines and Resources. Safeguards Policy Statement. December 20, 1974.

83. Jamieson, Don. Secretary of State for External Affairs. Notes for a Statement on Motions before the House of Commons. December 22, 1976.

84. Keely, James F., Autumn 1980. "Canadian Nuclear Export Policy and the Problems of Proliferation." *Canadian Public Policy—Analyse de Politiques,* VI, no. 4: pp. 614–627.

85. Gillespie, Alastair. Minister of Energy, Mines and Resources. Letter to the Minister of Energy, Mines and Resources Canada on the pricing of uranium exports. March 17, 1977.

86. *British North America Act, 1930.* [Revised Statutes of Canada (1970): 21 George V, c.26]

87. *Atomic Energy Control Act, 1946.* [Revised Statutes of Canada (1970): c.A–19, s.17]

88. See Bill C-14, the House of Commons of Canada, Third Session, Thirtieth Parliament, 26 Elizabeth II, November 1977; and Bill C-64, the House of Com-

mons of Canada, Third Session, Thirtieth Parliament, 26–27 Elizabeth II, June 29, 1978.

89. Bayda, E.D. 1978. *Cluff Lake Board of Inquiry: Final Report.* Regina, Canada: Government of Saskatchewan.
90. Greenwood, Ted. 1980. "Uranium." *Natural Resources in United States–Canada Relations. Patterns and Trends in Resource Supplies and Policies,* edited by Carl E. Beigie and Alfred O. Hero, Jr. Vol. II. pp. 319–393. Boulder, Colorado: Westview Press.
91. Bushnell, S.I. 1980. "The Control of Natural Resources through the Trade and Commerce Power and Proprietary Rights." *Canadian Public Policy–Analyse de Politiques.* VI:2.
92. Sproule, Kevin. 1979. "The Uranium Mining Industry in Saskatchewan: Control, Regulation and Related Constitutional Issues." *Saskatchewan Law Review.* Saskatchewan, Canada: University of Saskatchewan, College of Law. Vol. 43: pp. 65–79.
93. Elliot, William M. 1979. "Jurisdictional Dilemmas in Resource Industries." *Alberta Law Review.* Edmonton, Alberta: University of Alberta, Faculty of Law. Vol. XVII.
94. Mackintosh, Murray F. 1975. "Mineral Taxation." *Alberta Law Review.* Edmonton, Alberta: University of Alberta, Faculty of Law. Vol. XIII.
95. Wonder, Edward F. 1980. "Energy Bargaining in North America: Oil and Gas in Canadian–American Relations." In *International Energy Policy,* edited by Robert M. Lawrence and Martin O. Heisler. Lexington, Mass.: Lexington Books.

6 THE AFRICAN PRODUCERS

Uranium was first discovered in Africa in 1915 in what is now Zaire, formerly the Belgian Congo. Moderate quantities of ore were mined in subsequent years to recover radium, but by 1936 new discoveries—at Great Bear Lake in Canada—had all but eliminated demand for radium from Africa. It was not until after World War II that there was a major effort to discover and develop other uranium reserves in Africa. These efforts came first in southern Africa in response to the weapons needs of the United Kingdom and the United States, and then in francophone west Africa in response to the weapons needs of France.

In developing its weapons program, the United States depended first on the Belgian Congo, with imports via Belgium (where concentrate was refined) being recorded as early as 1937. In South Africa, large quantities of uranium were found associated with gold in the reefs of the Witwatersrand. Under agreement with the Combined Development Agency, uranium production began in South Africa in 1952, rising rapidly to a peak of nearly 5,000 MTU in 1959. Exploration efforts in Gabon and Niger were rewarded in 1956 and 1965 respectively, though production did not begin in Gabon until 1961 and in Niger until 1971. Weapons procurements by the United States and the United Kingdom declined in the mid-1960s and so too did South African uranium output. This production history is shown in Figure 6-1.

The beginnings of commercial nuclear power fuel demand began to be felt in the late 1960s, though much of the opening market, which was denied to non-U.S. producers by an embargo, was in the United States. However,

171

Figure 6–1.   Uranium Production In Africa.

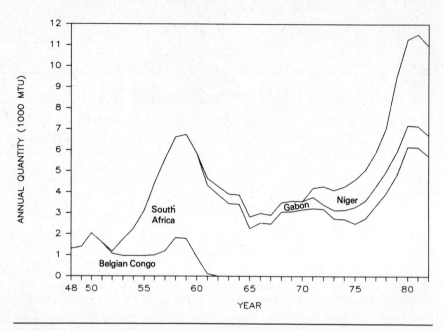

Source: Table B–1, Appendix B, and author's estimates for Gabon prior to 1969.

Gabon and Niger were able to increase exports to France during this period. It was not until the mid-1970s that the effects of commercial demand began to be felt seriously. Uranium market developments and advances in mining technology, together with economic pressures from South Africa, led the Rio Tinto Zinc corporation to develop a major deposit in Namibia, where output began in 1976.

Today, there are four major uranium-producing countries in Africa: Gabon, Niger, Namibia, and South Africa. Together these nations produced 48 percent of the (non-U.S.) world's uranium in 1982—a total of 14,840 MTU. The location of current and past production centers in these and other African nations are shown in Figure 6–2. Though there are many joint financing and other ventures, the uranium industry in Africa is dominated by France's Cogema (Gabon and Niger), Britain's Rio Tinto Zinc (Namibia), major South African gold companies, and the marketing agent Nufcor (South Africa). Politically, these four producers are quite different: Gabon is a one-party state; Niger is ruled by a military leader who gained power through a coup; Namibia is a white-ruled state dominated by South Africa but moving very slowly toward independence; and South Africa—also white-dominated—is struggling with black demands for political and economic power. What these nations all have in common is the potential for longer-term political instability, a fact that has a major influence on their role in the world uranium market.

As documented in earlier chapters, that market has undergone a series of radical changes over the past decade. From an era in the mid-1970s when producers had primary control of the market and prices rose well above production costs for virtually all African producers, new demand for

Figure 6–2.   Uranium in Africa.

Source: OECD Nuclear Energy Agency and the International Atomic Energy Agency. 1982. *Uranium Resources, Production and Demand*. Paris, France: Organization for Economic Cooperation and Development.

uranium has all but evaporated as nuclear power growth ambitions have been frustrated worldwide. Spot prices have dropped much closer to African production costs, and below in some cases, especially when correction is made for inflation and mining cost increases. How the major producers of Africa have responded to this changing market environment, and how their roles in the international market relate to domestic political and economic factors, is the subject of this chapter. In the following sections we explore the situation in each of the four countries in depth.

For the future, the role of the African producers in the international market must necessarily be a subject for speculation. The discovery of large low-cost deposits in Canada and Australia has increased the potential for price competition in a soft market, though the possibility of such competition will depend critically on government policies in those nations. More importantly, both Canada and Australia are likely to be regarded as potentially more stable long-term sources of supplies. That is, consumers may be more willing to enter into long-term supply arrangements with these nations than with the African states, which may thus have to compete—at best—in a depressed spot market. Such a contracting strategy is historically evident in South Africa, where delivery commitments have often been for only a few years. In contrast, similar arrangements in Canada and Australia may be for ten or more years.

This has two implications. First, the African producers are more likely to emphasize short-term or spot sales as the basis of their marketing strategy. Second, current production capacity will rapidly come out from under existing contracts. While official announcements indicate that South Africa and some of the other African producers are not eager to make sales at today's spot prices, such announcements may also reflect market realities and the producers' resolve may well be undermined in the next few years.

If utilities return to a spot-purchase acquisition strategy, the African suppliers may be a major competitive force in the market. Such a pattern would be a return to that of the early 1970s and an elaboration of the current pattern, which often involves spot purchases in a secondary market. But if utilities decide to put long-term supply "security" ahead of price—or if they can consistently get price concessions from Canada or Australia— the African producers will have a smaller share of the future market. This question will be decided first in the United States, where utility requirements generally come uncovered several years ahead of those of most foreign utilities.

SOUTH AFRICA

Most of South Africa's uranium is a byproduct from mines established many decades ago to produce gold. The production of uranium began in

1952, in response to the atomic weapons needs of the United States and Great Britain. A total of 43,913 MTU was produced between 1952 and 1965, with a peak of 4,954 MTU in 1959. Production had fallen to 2,262 MTU by 1965. After some recovery, production again declined in the early 1970s, but then soared with the boom in uranium prices, the 1959 peak output level finally being surpassed in 1980. However, the subsequent collapse of uranium prices, cutbacks in output, and abandonment of expansion plans strongly suggest that the early 1980s may have seen peak production for the decade.

Although coproduction with gold reduces the costs of recovering South Africa's uranium, it also has an unusual side effect: as the price of gold goes up, the production of uranium is reduced. Gold producers vary the ore grade which they mine with gold prices in order to extend the lifetimes of the mines. As the price of gold goes up, the producers mine lower grades of ore that would otherwise be less profitable, reserving the higher grades for times when gold prices are lower. Since uranium and gold in the ore occur in a fairly constant ratio, uranium output thus rises and declines in an inverse relation with gold prices.

Industry Structure

To date, uranium production in South Africa has come almost entirely from gold mines or from the slimes that have been produced in the past from such mines. An exception is the Palabora mine, operated by the Rio Tinto Zinc group, from which uranium is coproduced with copper. The few mines for which uranium is the primary product—and gold considered the byproduct—are experiencing financial difficulties. Most uranium and gold production is under the control of major South African mining houses. A list of these companies and the mines they operate is given in Table 6–1.

Byproduct Producers.   The difference between uranium as a byproduct of gold production and the coproduction of the two minerals is definitional. Only Harmony might be considered a coproducer, simply because it bases its reserve statistics on joint gold-uranium values. This distinction will be ignored here and companies producing both uranium and gold will be considered byproduct producers. In 1983, nine gold mines produced uranium as a byproduct, accounting for about 70 percent of total South African uranium production in that year.

The production of uranium entails relatively low costs for gold producers—costs have been estimated at about $12 per pound of U_3O_8.[1] Further, producing two economically unrelated minerals provides some protection against the cyclical nature of the minerals trade. High uranium profits during the period of low gold prices (1976–1979) boosted the overall profitability of many of these mines, and the subsequent high price of gold helped to off-set slumping uranium profits. However, few of the mines appear to depend critically on uranium profits.

The level of a mine's uranium operations depends on its contractual commitments. Since costs are low, most existing contracts are profitable; but where production is uncommitted, uranium operations are at risk. Since 1982, high South African interest rates have made stockpiling expensive and a general uranium surplus has made spot sales—always important in South Africa's trade—more difficult. Gold profits have been used to subsidize uranium stockpiling, which was about 850 MTU in 1979 and estimated to be 1,000–1,300 MTU in 1980.[1] Subsequently, as can be seen in Table 6-1, a number of mines reduced their uranium production to diminish the need for stockpiling.[2] The need for contracts to justify uranium mine operations has resulted in some companies and mines shutting down while others expand.

South African producers fear both the weakness of the near-term market and potential long-term competition from a number of large, foreign, low-cost high-grade producers.[3-6] In contrast to those of some other nations, many South African contracts have only a few years to run. Given the recent expansion of world supply potential and a continuing downward revision of nuclear power growth estimates, the concern over long-term competition seems justified.

South African mining companies responded to favorable market conditions in the mid-1970s with a wave of expansion that, in some cases, is only now being completed. Vaal Reefs expanded to 1,520 MTU per year of capacity, having been very successful in its uranium sales activities, but by 1982 was reducing production to match sales.[7] A new uranium extraction plant was built at the Western Areas Gold Mine (operating in the Middle Elsberg Reefs) that began operation in late 1982.

Operations at Randfontein expanded further with the mine being rehabilitated and re-opened in 1978 when gold and uranium prices went up and a sale was concluded for 770 MTU per year. However, ore grades at Randfontein have been much lower than anticipated. As a result production is only half of the level planned, requiring Randfontein to purchase from other producers to meet its contractual commitments. A new mine shaft should raise production capacity output to 1,060 MTU per year by mid-1984.[8]

Other producers are experiencing marketing problems. Western Deep Levels is having trouble finding buyers and may cut back production. Hartebeestfontein has stopped production from slimes and is relying on production from freshly milled ore to meet deliveries. This results in a small reduction in output but yields a larger reduction in costs. The head of Rand Mines' gold and uranium division has indicated that uranium production at Blyvooruitzicht might be curtailed without an improvement in spot prices,

Table 6-1. Production Capacity of South African Producers (*MTU Annually*).

Group Mine/Company	Recoverable Reserves	Production					Production Capacity		
		1978	1979	1980	1981	1982	1983	1984	1985
Gold Fields of South Africa									
W. Driefontein	34,000	250	238	213	280	190	280	280	280
Anglo-American									
Afrikander Lease	13,000	0	0	115	(169)	(254)	(340)	(340)	(340)
Vaal Reefs	110,000	889	1,152	1,490	1,525	1,459	1,525	1,525	1,525
Western Deeps	16,000	155	165	180	186	155	186	186	186
ERGO	3,000	100	129	250	250	224	250	250	250
JMS	17,000	553	575	828	850	732	850	1,000	1,000
Barlow Rand									
Blyvooruitzicht	7,000	248	236	275	280	214	280	125	0
Harmony	22,000	454	447	416	560	501	770	770	770
Johannesburg Consolidated Investments									
Randfontein	100,000	81	345	548	410	392	410	610	810
Western Areas	13,000	0	0	0	0	145	340	420	510
General Mining Union Corporation									
Buffelsfontein	25,000	526	563	512	580	492	580	580	580
Chemwes	13,000	0	205	590	590	514	590	720	720
W. Rand Con.	2,000	265	340	327	0	0	(340)	(340)	(340)
Anglo-Transvaal									
Hartebeestfontein	20,000	310	325	370	380	364	380	380	380
Palabora	n.a.	119	96	100	100	100	100	100	100
Beisa	13,000	0	0	0	0	215	385	385	650
Totals:	408,000	3,950	4,816	6,124	5,991	5,697	6,641	7,386	7,861

Sources: OECD Nuclear Energy Agency, and the International Atomic Energy Agency. *Uranium: Resources, Production and Demand.* 1979. Paris, France: Organization for Economic Cooperation and Development; "Analysis of Rand and O.F.S. Quarterlies," Supplement to *Mining Journal* (London), various issues.

which, at \$17 per pound of U_3O_8, he described as dangerously close to break-even for a number of mines.[9]

New uranium development in South Africa is being postponed indefinitely, even where new gold production is being planned. Anglo-American has announced that its new Erfdeel gold mine near Free State Saaiplaas, scheduled to begin production in 1986, will not produce uranium despite the presence of good ore grades. The mine will instead stockpile tailings for possible future production.[1]

Uranium from Slimes. Some South African uranium is recovered from "slime dams," large piles of tailings from earlier gold mining, which can be processed for uranium the same way virgin ore can. In fact, advanced technology allows gold to be recovered from the "depleted" ores as well. At present, the Orange Free State Joint Metallurgical Scheme (JMS), the East Rand Gold and Uranium Operation (ERGO), and Chemwes Ltd. are producing uranium from slimes. Together, these companies produced a peak of more than 1,600 MTU in 1981.[10]

The Joint Metallurgical Scheme (JMS) operation involves a revival and expansion of the President Brand uranium plant, which was mothballed in 1971 due to the weak market. JMS now processes slimes from Free State Geduld, F.S. Saaiplaas, President Brand, President Steyn, Welkom, and Western Holdings—all part of the Anglo-American group, which owns the plant (see Table 6–2). The plant was expanded so that by early 1980 it could process 500,000 tons of slimes per month, versus an original design capacity of 180,000 tons per month. Uranium production peaked in 1981 at 850 MTU, then declined in 1982 due to declining ore grades. Production in early 1983 was substantially lower for the same reason. As a result, the value of the gold produced now approaches that of the uranium. Gold ore grades have fallen slightly, but higher throughput has allowed production to remain close to 4,000 kilograms per year.

ERGO is also an Anglo-American subsidiary that coproduces gold and uranium from the tailings of abandoned slime dams. Full production was reached during 1979, and 250 MTU were produced the following year.[10] Production fell 13 percent in fiscal year 1982 due to smaller amounts of slime being treated, and a further decline of 24 percent is expected in fiscal year 1983, to 170 MTU, due to declining ore grades.[11] ERGO also produces 7,000 kilograms of gold a year, which even at current prices is far more valuable than the uranium produced. Expansion to the slime dams on the Far East Rand was deferred due to the relative abundance of uranium and sulphur over gold. ERGO is instead conducting talks with East Daggafontein Mines on the feasibility of a joint operation to recover gold from their tailings, with uranium and sulphur to be recovered at a later date.[12]

The Chemwes, Ltd. operation is a subsidiary of General Mining/Union

Table 6–2. Producers of Gold and Uranium.

Group	Mine/Company	Uranium Profit (Thousand Rands) Fiscal Year Current	Previous	Type of Producer	Ore Grade (%)
Gold Fields of South Africa	W. Driefontein	9,029	6,389	Byproduct	.016
Anglo-American	Vaal Reefs[a]	30,289	42,534	Byproduct	.023
	Western Deep[a] Levels	2,214	5,072	Byproduct	.09
	East Rand Gold[b] and Uranium (ERGO)	40,309	27,944	Slimes-Byproduct	.0014
	Joint Metallurgical Scheme +	28,656	23,893	Slimes-Byproduct	.016
Barlow Rand	Blyvooruitzicht	11,886	5,375	Byproduct	.013
	Harmony[c]	42,631	32,246	Coproduct	.09
Johannesburg Consolidated Investments	Randfontein[a]	19,166	12,034	Byproduct	.017
	Western Areas	2,925		Byproduct	.041
General Mining/Union Corporation	Beisa	n.a.	n.a.	Primary	.018
	Buffelsfontein	6,424	14,804	Byproduct	.013
	Chemwes[a]	35,579	42,193	Slime-primary	
Anglo-Transvaal	Hartebeestfontein	13,055	10,871	Byrproduct	.014

Notes:
a. Current fiscal year represents 12 months to 12/31/82; for all other mines, 9 months to 3/31/83 shown for current fiscal year, versus same period in previous fiscal year. ERGO shows 12 month results to 3/31/83, and JMS represents 6 months results to 3/31/83.
b. Uranium and sulphuric acid profits.
c. Uranium and pyrite revenue.

Source: *Mining Journal*. April 29, 1983 Supplement. London: Mining Journal Ltd.

Corporation (Gencor), and is owned by the Buffelsfontein and Stilfontein mines. Unlike ERGO, it primarily produces uranium, with the output split between the owners according to equity share—85 percent to Stilfontein and 15 percent to Buffelsfontein. This unequal sharing is a remnant of the quota system used in the 1960s, when the Combined Development Agency (CDA), discussed in Chapter 2, found itself with excessive supplies and sought to stretch out the deliveries over time. However, the quotas assigned to some producers were too small to allow economic production, so they sold their "quotas" to other mines that could then produce at a comfortable level. Stilfontein sold its quota to Buffelsfontein and stockpiled its own uranium-bearing tailings. Subsequently, in 1978, Stilfontein decided to develop a plant to process these slimes, along with some from Buffelsfontein. With the high-grade tailings from Buffelsfontein now depleted, the only material still being processed is from Stilfontein.[13] The particular administrative arrangement results in part from the tax advantages of operating a chemical operation, rather than a gold mine.[14]

Until 1981, slimes were also being processed for uranium extraction at the Hartebeestfontein mine, part of the Anglo-Transvaal group, but production of uranium from tailings was then replaced by extraction of uranium from freshly mined ore at lower cost.[15] The impact of this decision on production levels was small.

Primary Producers.   Production of uranium in South Africa as a primary product (instead of as a secondary product to gold) is not very successful in today's market. In the gold fields, there are three mines whose primary product is uranium: one operating, another that has its ore treated elsewhere, and a third that stopped production in 1980. Non-associated uranium has been discovered at several sites outside of the gold fields. While there are no current plans to pursue production from these deposits, they clearly represent a source of long-term uranium supply.

Afrikander Lease is a primary producer that has leased its mineral rights to Vaal Reefs, the country's largest uranium producer. According to an arrangement worked out between the two companies in 1979, Vaal Reefs processed approximately 15,000 tons of ore per month from Afrikander Lease, for which Vaal Reefs agreed to pay a royalty of 5 percent on revenue from sales, plus additional payments if profits exceed 30 percent of revenues, and to provide financing for the new mine. This has substantial tax advantages for both sides. A new plant designed to process 50,000 tons of ore per month was being built to come on line in 1982, but after $180 million of expenditures, it was mothballed to await better uranium market conditions. Vaal Reefs was unable to sell all its production in 1980 and, with the suspension of the new Afrikander Lease capacity, Vaal Reefs will use its own lower-cost resources to meet Afrikander Lease's sales commitments, including one contract extending from the end of 1984 to 1990.[16]

Gencor's Beisa mine began operation as South Africa's second primary uranium producer early in 1982.[17] Beisa was bought out in 1982 by St. Helena, a gold mine also in the Gencor group, in part to provide a tax write-off for St. Helena.[18] Its chances for success should be better than those of Afrikander Lease for two reasons. First, it has contracted sales for a substantial amount of its output at what are believed to be favorable terms.[17,19] Second, its costs appear to be low, with profits ensured at $400 per ounce of gold and $12 per pound of uranium, assuming at least 80 percent of the uranium is sold.[20] Production in 1982 was 215 MTU, with capacity reaching 385 MTU annually by early 1983. Initial uranium production will be stockpiled until the contract takes effect in 1984.

West Rand Consolidated had long been considered a primary uranium producer because its gold production costs were so high that its profits came mainly from uranium sales. In fact, most of that profit was derived from spot sales of uranium; its long-term contracts, which it had never been able to renegotiate,[17,21] were in the $8 to $10 per pound range. When the spot market softened, reported profits plummeted, and the mine directors voted to stop production of uranium as of August 1982.[22]

The only other prospect for primary uranium production in South Africa at present is near Beaufort West and Sutherland in the Karoo area of Cape Province, where exploration has discovered some small deposits. Esso Minerals and Union Carbide have both announced discoveries, but neither has revealed any plans for production. At the same time, Randfontein and Johannesburg Consolidated Industries (JCI), exploring the area in conjunction, have found several small deposits but are not planning to exploit them except as satellites to a larger deposit, should one be found.[23,24]

Uranium Trade

With the exception of Palabora, South African uranium is marketed through Nufcor, a private company managed by the country's major uranium producers. Nufcor also re-refines uranium to ensure an export product that is uniform in grade.

Initially, Nufcor marketed output from mines that were still producing after weapons purchases ended. In this activity it was apparently successful, since most mines reportedly had only small stocks remaining by 1977.[25] Unfortunately, South Africa's Atomic Energy Act forbids the release of details concerning uranium contracts, including quantity or customer. It is thus difficult to determine or confirm delivery schedules or contract terms, except through consumers, for early years. However, as mines became more aggressive in their sales activity in the mid-1970s, with many contracting relationships also involving consumer investment or a direct tie between contracts and expansion investments, more information has been released, so that a picture of marketing strategies has begun to emerge.

In the buyer's market of the late 1960s and early 1970s, Nufcor apparently sold what it could at the prevailing prices. Unfortunately for producers, a number of long-term contracts were written that set prices at fixed levels of $8 to $10 per pound of U_3O_8. While this may have been profitable for a South African producer in the early 1970s, inflation and rising real costs made these contracts quite unprofitable by the end of the decade. Many producers renegotiated their contracts in the seller's market of the late 1970s, but several could not. However, with the price of uranium soaring by a factor of five in the course of two years, even disadvantaged producers were able to make spot sales that were extremely profitable.

As the market tightened and prices began to climb, South Africa's producers took advantage of the seller's market to write new contracts that included consumer financing of plant and mine expansion, usually interest-free, although with lower-than-market prices for the uranium output that consumers would ultimately receive in return. As in other nations, expansion would not be undertaken without specific sales lined up. Often the loan would be designated for the plant that was to provide the uranium and assigned to the mine's capital expansion account, rather than being entered as revenue from sales, in order to reduce taxes. Many contracts also indexed prices, the index usually being based on mining costs in the gold industry.

By the end of the decade, most mines had signed contracts covering a significant part of their production into the midterm future, often with prepayment in the form of consumer loans. However, it appears that producing companies did not commit all of their production in long-term contracts. They left a margin that allowed for production problems, including the exploitation of lower grades of ore as the price of gold rose. Moreover, many mining companies historically made significant profits from spot sales, suggesting a desire to maintain a capability to enter the spot market. On average, the contracts written in the 1970s appear to cover about 80 percent of projected production, with the remainder allocated for spot sales. If spot sales cannot be made, uranium can be stockpiled as yellowcake or production of uranium reduced and the uranium, in effect, stockpiled in tailings for later recovery.

Although it is not possible to document all of South Africa's sales, data available from customers, government agencies, and mining and uranium industry publications have allowed us to develop estimates of exports by destination. Such estimates are shown in Figure 6-3. Because of extreme secrecy and the fact that significant quantities have been exported under hard-to-document spot sales, the export profile shown in Figure 6-3 should be considered as minimum export levels. For example, production in 1979 was 4,700 MTU, and 850 MTU were reported stockpiled;[1] our export figure for that year of 2,650 MTU thus seems to be low by about 1,200 MTU. Similarly, the 1980 numbers appear to be low by 1,215–1,470 MTU.

The earliest commercial customers for South African uranium were the Japanese, with most of the quantities shown in Figure 6–3 contracted by the early 1970s. In the years since, the Japanese have sought to diversify new supplies among other producers (including Australia, Namibia, and Niger) for deliveries extending over the remainder of the decade. In addition, delays in nuclear power programs have resulted in greater caution in making new forward commitments. Correspondingly, Japanese utilities appear not to have pursued additional contracting with South African suppliers. Only Beisa is known to have signed a long-term contract with Japanese buyers in recent years. The price negotiated by the Japanese with Nufcor for 1984 deliveries reportedly is $33.50 per pound of U_3O_8, down slightly from 1983 prices.[26]

German consumers, on the other hand, were slower to cover their needs and did not enter into major supply arrangements until the mid 1970s. At this time, German uranium companies and utilities entered into major supply commitments from South Africa for deliveries then and into the 1980s. Because German organizations have shown a greater propensity for spot purchases than most other consumers, it is reasonable to assume that West Germany is often the unidentified customer in many reports of spot sales.

Consumers from a number of other nations have been involved in uranium trade with South Africa. At least one major sale was made (by Randfontein) to France, and several to U.S. reactor vendors and fuel fabricators. Switzerland, Belgium, and Austria have purchased comparatively small quantities, but of these only Belgium's contract now extends any distance into the future. Uranium sold to Austria was enriched in the Soviet Union and later resold to a U.S. utility. Spain appears to be scheduled to receive significant quantities through this decade. The most recent long-term sale reported was to Taiwan in 1982. This will very likely prove to be one of the first of a series of independent purchases by third world nations no longer depending on the United States or other nuclear technology-exporting nations.

It is evident from Figure 6–3 that a significant fraction of known South African production in the past is unaccounted for. This suggests either a highly successful effort at secrecy or, more likely, a large component of spot sales that have not been reported. It is also evident that South Africa's known foreward-export commitments do not extend very far into the future, nor do they account for more than a small fraction of future production potential. This is in strong contrast to the long-term supply arrangements entered into by Canada and Australia (and even some other African producers). South Africa may thus be an important source of spot sales of uranium in the future as well as in the past, though—as we shall see in the next chapter—there are few consumers outside the United States that will need significant additional quantities of uranium during the remainder of this decade.

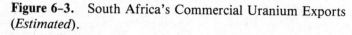

Figure 6-3. South Africa's Commercial Uranium Exports (*Estimated*).

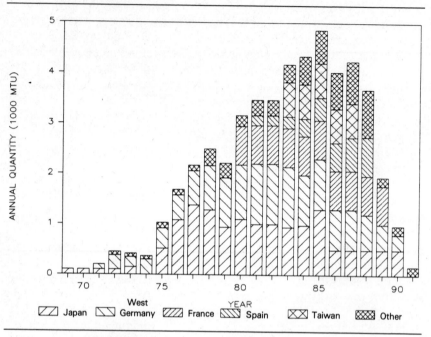

Note: Estimates may understate actual exports due to the high degree of secrecy associated with South African exports and because it is difficult to identify spot sales.

Policy Context

When supplies of a commodity are abundant, it is not unusual for buyers to pay increased attention to nonprice considerations. In the case of South Africa, this could include concern about long-term supply stability as well as the stigma attached by some to trade with South Africa. Historically, Nufcor has been an exceptionally reliable supplier, especially when compared with the delays and interruptions to which other producers have been subject. Yet there is a continuing concern about future internal stability, which may be partly responsible for South Africa's apparently small number of long-term supply commitments.

Opposition to South Africa's apartheid system is common and has resulted in a degree of diplomatic isolation. Trade with South Africa has been routinely criticized, and OPEC has imposed an oil embargo on South Africa. However, the country experiences little trouble in commercial activity; at worst, embargoes and other actions simply increase costs to South

Africa. Moreover, many multinationals have invested in the country, taking advantage of low labor costs and the infrastructural benefits of the most developed economy in sub-Saharan Africa.

Since sales of uranium have been generally low since 1979, it is difficult to assess the effects of increasing opposition to apartheid as it affects sales of South African uranium. The threat of political opponents of apartheid in countries like the United States and Germany to use the regulatory process to embarrass those that import South African uranium may have deterred some utilities, especially where uranium is readily available at comparable prices from politically more acceptable producers. Whatever the effect, it would undoubtedly be less if the market were significantly tighter or if South Africa's prices were substantially better than those of other producers.

While consumers' decisions may be directly affected by apartheid, it is more likely that South Africa's uranium exports will be affected indirectly, through threats to the political stability of the country (and external perceptions of stability), and through the labor situation.

Since World War II, the National Party has dominated political life in South Africa, but there are increasing signs of change. Pressures from white liberals, "Coloureds" (people of mixed race), and Asians, as well as blacks, have all influenced the regime to make concessions and liberalize the political system. However, these moves have met with resistance from white conservatives within the party. As a result, the prime minister has found himself talking of liberalization while carrying out policies (or enforcing policies which were rarely enforced) that often have the opposite effect. The most important step thus far undertaken has been the setting up of separate legislative assemblies for Asian and Coloureds, giving these groups limited powers and self-rule, while strengthening the power of the white parliament. This has made the regional black homelands the focus of whatever representation blacks are to be allowed. While these steps have altered the focus of opposition, they seem unlikely to satisfy the objectives and desires of the non-white majority in the long term.

In the near term, South African apartheid creates two kinds of problems of specific concern to the mining industry: it increases general political instability and it creates labor problems. Because the apartheid system maintains that all blacks are natives of "homelands"—rural, undeveloped areas demarcated by the South African goverment—mine workers are legally considered transients and are allowed only temporary work permits. There are also limits on accommodations for married blacks to 3 percent of the labor force, and restrictions have been placed on the types of jobs available to blacks, partly to keep them below certain skill levels and partly to keep them out of "sensitive" jobs, such as those involving explosives. In these ways, apartheid has aggravated a severe shortage of skilled labor in the mining industry, a situation that can be expected to worsen given the limited white labor available. As a result, the industry is being forced to push for reform of the country's labor laws in order to meet long-term goals, especially where expansion is planned.[27]

To compensate for these conditions, the industry has raised wage scales for blacks at a higher rate than whites—tenfold over the last ten years. In 1982, blacks received a 16 percent average annual wage increase, versus about 14 percent for whites. The gap between white and non-white wages, though still large, is thus narrowing, and some see a future with a non-racial wage structure.[28] Mining companies also increased the number of "career men," that is, workers who continue on a job year after year.[29] However, a white backlash to this "preferential" treatment for blacks is possible: white labor unions have traditionally been the strongest opponents of racial reform.

Although government-sponsored analyses indicate that relying on the white population even for skilled manpower would prevent South Africa from attaining its full development potential, government labor policy is still unclear, and racial policy is, if anything, increasingly restrictive.[28,30] This contradictory situation will undoubtedly pose a difficult problem for the South African mining industry.

Since uranium output has declined, the effect of labor restrictions might seem to be lessened. However, expansion of gold production will probably result in escalating labor expenses, which already account for more than 50 percent of costs in gold mining.[31] With most capital expenditure already made, high inflation rates, and increases in wages to blacks for salary equalization, this proportion can be expected to grow.

As the only voice of black grievances, the black unions are increasing in power and their demands are becoming more far-ranging. Thus far, labor unrest has been primarily confined to manufacturing industries, but the nation's underlying racial conflicts may spread to the mining industry, where it will invariable cause problems for uranium producers. In recent years, black mine workers have rioted over such diverse issues as proposed changes in death benefits (at President Steyn), dormitory rent increases in Johannesburg, and safety-related matters.[32]

While such problems seem likely to disrupt uranium and gold operations at some future time, it is not obvious that these disruptions would seriously undermine reliability of supplies from South Africa. The importance of gold mining to the nation's economy, which might make it an obvious target of opponents to the current regime, also suggests that major efforts would be made to resolve such problems. What may be most affected is the nature of uranium trade with South Africa. Given major uncertainties, as well as anticipated market conditions, it seems likely that consumers will come to South Africa—as they have in the past—for spot and short-term purchases of uranium, rather than for long-term supply security.

NAMIBIA

Uranium mining and exploration activity in Namibia centers around the Rössing mine, located near Swakopmund, inland from the central coast. Although uranium mineralization was discovered at Rössing more than fifty years ago, the low grade of the ore made exploitation uneconomical until 1966, when Rio Tinto Zinc (RTZ), the London-based minerals conglomerate, planning to use extraction methods developed at their Palabora copper mine in South Africa, obtained the rights to exploit the deposit.[33] Initially, ownership included the government-mandated 50 percent "local" interests, namely South Africa's General Mining and the Industrial Development Corporation,[34-36] but as the requirement was eased to 25 percent in 1973, part of this share was sold. RTZ reduced its holding as well, transferring some ownership to its subsidiary Rio Algom of Canada, in order to obtain additional financing. The French firm Minatome was also brought in for this reason.[37] Table 6-3 shows what is publicly known about the current ownership of Rössing. The initial funding reportedly came from loans obtained by RTZ (80%), General Mining (10%), and Urangesellschaft (10%),[38] which had been repaid by the end of 1982. The South African tax exemption, which had been granted Rössing, ended at the same time, and with taxes anticipated to be about R30 million, after-tax profitability may decrease.

Attempts at commercial production at Rössing began in 1975, but problems with abrasive ore that damaged machinery and a fire that destroyed one facility in 1978 caused delays in the actual startup of the mine. Still, it produced 3,840 MTU in 1979 and reached its full[39] annual capacity of 4,230 MTU by the end of that year, although changing ore grades and other factors will cause actual output levels to vary. While the deposit is very large, with an estimated 300 million tons of ore, the ore grade is low, on the order of 0.03 to 0.05 percent.[40] However, economies of scale, combined with the low-acid feed required to process this particular type of ore, serve to reduce unit costs.

Based on reports of sales of $291 million in 1979, when production was 3,840 MTU,[39] the average sales price in that year appears to have been $29 per pound. Based on profit and revenue reports, production costs seem to have been about $17 per pound in 1978 and $22 per pound in 1979. Although final sales and production figures are not available for 1980, RTZ has reported that profits from Rössing increased by just under $3 per pound in 1980, due to higher prices, greater operating efficiencies, and higher production.[41] The reason for the apparent increase in costs from 1978 to 1979 is not publicly known, but may include escalating labor costs and replacement of the plant destroyed by fire. In addition, changing currency relationships (among the rand, the pound, and the dollar) have affected dollar-denominated cost reports.

Table 6-3.   Rössing Ownership Shares.[a]

Rio Tinto Zinc	41.35 percent
Rio Algom (Canada)[b]	10.0 percent
General Mining (South Africa)	2.3 percent
Industrial Development Corp. (S.A.)	13.47 percent
Minatome (France)	10.0 percent
Others	22.88 percent

Notes:

a. Note that nearly one-quarter of the mine is owned by unknown partners. Note also that the South African share is only 16%, less than the 25% required in 1973 by the South African Minister of Mines. While the requirement may no longer be applicable, it may well be that some of the unidentified equity belongs to South Africans. Urangesellshaft is also believed to have an equity share and it has been suggested that Iran held shares in the mine.

b. Subsidiary of RTZ.

Sources: Rio Tinto Zinc and Rio Algom's shares are taken from the 1978 RTZ Annual Report. The other figures are the author's estimate based on slightly conflicting data from a variety of sources.

As can be seen in Table 6-4, sales, profits, and costs fell in 1981, then rose in 1982, in part due to currency fluctuations. Lower ore grades were partially to blame for reduced production, though the amount of ore processed also fell by 4.5 percent.[42] Production is expected to fall slightly in 1983 as well. Sales and profits have continued to rise. In part, this is the result of the rise of the dollar against the rand and the pound, since Rössing costs are paid in the latter currencies while income comes from dollar-denominated sales. However, production costs appear to have increased in the past few years, at a rate that would make new sales barely profitable, as indicated in Table 6-4.

Rössing's contracts were signed in the early 1970s, when prices were very low. These contracts were subsequently renegotiated based on at least two different formulas. The contract with Japanese utilities was revised to a market formula that reportedly brought Rössing's price to $32 per pound in

Table 6-4.   Rössing Income and Cost Statistics.

	1978	1979	1980	1981	1982
Production (MTU)	2,697	3,840	4,042	3,970	3,780
Sales ($ million)	130	291	n.a.	256	353
Profits to RTZ ($ million)	4	27.2	45.5	256	353
Apparent price ($/lb. U_3O_8)	18.5	29.1	—	34.6	52.3
Apparent cost of production ($/lb. U_3O_8)	17.3	22.5	(lower)	20.1	23.2

Sources: News Agencies May 14, 1980. Cited in *Africa Research Bulletin*, June 1980; *Nuclear Fuel*. May 28, 1979 and April 27, 1981; OECD Nuclear Energy Agency, and the International Atomic Energy Agency. *Uranium: Resources, Production and Demand*. Paris, France: Organization for Economic Cooperation and Development. p. 130; *Economic Intelligence Unit*. 1981. pp. 35-36. *Nuclear Fuel*. May 9, 1983.

1978, or about $10 per pound below spot-market prices at that time. The British contract was changed to a cost-plus-profit formula[37] with deliveries in 1983 and 1984 said to be at about $34 per pound of U_3O_8. According to other reports, Japanese utilities paid $37 a pound in 1978, $43 in 1979,[43] about $45 in 1981, $38 in 1982, and $35.25 in 1983, with $34.25 negotiated for 1984.[26]

Three promising deposits of uranium-bearing ore have been discovered in the vicinity of Rössing. General Mining, a South African company with an interest in Rössing, has delineated a deposit at Langer Heinrich. Although South Africa's Atomic Energy Act bars the release of information about the discovery, the deposit appears to be substantial, though not as large as Rössing, with twice the ore grade. The nearness of the deposit to Rössing and its established infrastructure enhances its value. A pilot-plant test begun in 1977 has been successfully completed, and plans have developed for extraction of about 1,000 MTU per year. However, further commitments seem very unlikely, given political uncertainties about Namibia and the condition of the world uranium market.

The two largest gold-mining companies of South Africa have also reported finding signs of uranium mineralization in Namibia. Anglo-American, with the Union Corporation and the French companies CFP and Elf Aquitaine, have found a high-grade deposit to the west of Rössing.[44,45] Goldfields of South Africa has found a deposit at Trekkopje, to the northeast of Rössing.[46] In addition, Union Carbide, Falconbridge Nickel Mines, and Johannesburg Consolidated Investments have been active in uranium exploration in Namibia. Further exploration or development activity is almost certain to depend on resolution of political uncertainties about Namibia's future governance, as well as on improvement in uranium market conditions.

Uranium Trade

South Africa's Atomic Energy Act and the reticence of Rio Tinto Zinc have been fairly effective in preventing details of Namibian uranium contracts from emerging publicly. Only a few contracts are known with any certainty, although it is possible to infer several others. The two major early customers for Namibia's uranium were Britain, which was scheduled to receive 7,500 MTU from 1976 to 1982, and Japan, which was scheduled to receive 8,200 MTU from 1975 to 1984. Startup problems at Rössing led to some slippage in these deliveries.[47] Japan's political discomfort over trade with Namibia ultimately led to cancellation of the Japanese utility contracts for Namibian uranium, although contracts for a similar amount were instituted with the parent company, RTZ. A number of reports tie the Japanese contracts to RTZ's Namibian production, and it seems unlikely that RTZ could supply such large amounts without using Namibian production.[48] Britain, on the other hand, has resisted pressures to cancel its contracts, for which deliveries are to end in 1984. Late in 1982, Taipower contracted for 3,400 MTU from RTZ, with no origin specified. Rössing seems the most likely source of material for this contract, given the excess capacity that will be available.

In 1976, the chief executive of Rössing, A. Macmillan, stated that all output had already been sold, although the time period and production levels were not specified. Other customers have not confirmed the existence of their contracts. There are reports that Germany's Urangesellschaft, despite pressures to avoid financial involvement in the mine, did commit to purchase some uranium from Rössing.[49] France's Minatome appears to be receiving at least its equity share, and possibly more.[50] A number of additional contracts were signed with RTZ by Japanese firms, such that the quantities involved could only come from Rössing. In addition, repeated reference in the trade press was made to Iranian purchases of as much as 30,000 to 50,000 STU_3O_8,[51] and the chief executive of RTZ, Alistair Frame, referred publicly to deliveries under Iranian contracts. Subsequently, any such contracts were cancelled by the Khomeini regime, although at least one delivery was made (probably to the United Kingdom for conversion). Reports have indicated that 80 tons of uranium were being held in France for Iran, but the source of that uranium is not known.[52] Figure 6–4 presents the author's estimates of deliveries from Rössing, based on published and unpublished information.

Policy Context

Namibia is a large, sparsely populated country on the southwestern coast of Africa. It is potentially rich in mineral resources, though economic and political circumstances may significantly constrain exploitation. Namibia is currently under the control of South Africa, though this control is contested both by local groups and many nations, acting independently and through the United Nations.

Originally colonized by Germans in 1884, Namibia (then called South West Africa) was occupied by South African armies in World War I, and subsequently placed under a League of Nations mandate. Following World War II and the demise of the League of Nations, South Africa, contending that the mandate had expired, claimed unrestricted sovereignty over the territory. Since then, the occupation of Namibia has become the source of a major international dispute and a small-scale guerilla war. After attempting to incorporate the territory as a fifth province, South Africa yielded to growing international pressure and agreed to grant Namibia its independence.

For several years, Britain, West Germany, France, Canada, and the United States have been attempting to achieve an agreement between the

Figure 6-4. Namibia's Commercial Uranium Exports (*Estimated*).

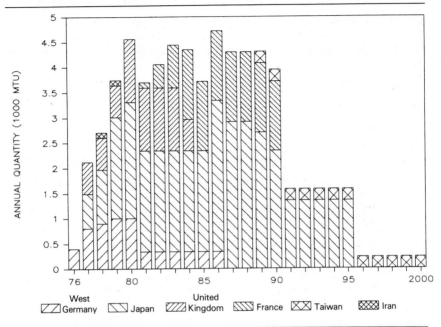

South West African People's Organization (SWAPO)—the movement fighting for Namibian independence—and South Africa and its negotiating arm, the Democratic Turnhalle Alliance (DTA), as to the nature of that independence. Although resolution of the conflict has appeared imminent at several points, new disagreements have always arisen, usually initiated by South Africa. Some diagreements, such as the size of a demilitarized zone or monitoring of SWAPO bases by U.N. observers, have been technical in nature. Others, such as constitutional guarantees of minorities' rights, have been more fundamental. The Reagan administration's insistence in 1982 on the removal of Cuban troops from Angola as part of a Namibian settlement further complicated negotiations and split the Western team for the first time. Resolution of the conflict over Namibia appears to be as distant as ever.

One reason that guarantees for minorities are deemed important in Namibia is the preponderance of one ethnic group, the Ovambo, and their strong representation in SWAPO. Currently, the Ovambo make up over half the population.[53] A second ethnic group, the Herero, who actively rebelled against the Germans during the early colonial days (and were nearly annihilated as a result), are more closely aligned with the present South

African-sponsored government, while the Ovambo have generally been dissatisfied with the existing regime. Thus the general political conflict over Namibia has interacted with ethnic tensions to raise the stakes in any resolution of the governance problem. Some are concerned that the Ovambo will dominate a post-independence Namibia, although their own unity is questionable since the Ovambo group is actually made up of a number of different tribes that have often warred with each other.[54]

The other major conflict that relates to independence concerns Namibia's only major port, Walvis Bay, which is claimed directly by South Africa and by SWAPO on behalf of Namibia. Walvis Bay was originally settled in 1876 as the Cape Colony, at that time a colonial possession of England. With the end of the Boer War, Cape Colony and Walvis Bay became part of South Africa, providing the basis for South Africa's claim to the area. In 1922, South Africa united the port with the rest of Namibia for administrative purposes, and this, along with its geographic proximity, provides the basis for SWAPO's claim. The contention that Walvis Bay is an essential part of Namibia is supported by the United Nations. In 1977, South Africa annexed Walvis Bay and the offshore Penguin Islands by special proclamation. Should Namibia achieve independence from South Africa without settlement of the Walvis Bay issue, disputes over this port could well continue, affecting uranium shipments from the area.

For a time in 1978, RTZ reportedly employed the French airline UTA (which serves west Africa) to transport uranium from Namibia to France.[55] The public rationale included RTZ's desire to avoid large penalties for late deliveries, and insignificant differences over sea-freight costs, but there was undoubtedly also a question of political sensitivity, with RTZ restricting shipments of uranium through South African seaports in order to mollify the potentially unfriendly new rulers of Namibia. However, French participation in this arrangement ended when it became public, and uranium once again was transported by sea.[56]

Namibia's economy depends almost entirely on mining. Other industries had been devoted to processing the fish catch, but overfishing has nearly eliminated these industries.[57] Fishing fleets were based in Walvis Bay and owned, for the most part, by South African concerns. Domestic manufacturing is discouraged by the ready flow of manufactured goods from South Africa. Most commercial agriculture is derived from livestock production in the south and center of the country, mostly cattle and karakul, which contribute about 10 percent of the country's gross domestic product (GDP) and 20 percent of its exports. A serious drought in 1982 and 1983 resulted in significant losses in this sector, and—together with civil unrest—resulted in up to half of the white-owned farms in the northern part of the country being abandoned.

The most recent data indicate that all mining operations combined contribute about 50 percent of the country's GDP and 70 percent of her exports.[58] The uranium industry is second only to diamonds, which contribute 30 percent of the country's GDP. Rössing now appears to be the second largest mineral exporter in terms of sales, a position previously held by Tsumeb Corporation, which produces and exports copper, lead, zinc, silver, cadmium, and tin.

Beginning in the late 1960s, interest among foreign companies in the minerals of Namibia increased greatly; this interest was boosted by a 1973 decision by the South African Minister of Mines to reduce required "local"; participation (i.e., South African involvement) from 50 to 25 percent. In addition, the tribal "homelands" were opened up to exploration. These moves brought a threefold increase in the granting of mining concessions, from 36 in 1969 to 104 in 1974. However, since the opening of the Otjihase copper mine in 1975 and Rössing in 1976, no new mines have been brought into production, although a number of mineral prospects are ready for development pending the settlement of Namibia's political future.[59] SWAPO has indicated that it welcomes foreign mineral producers, but would seek more equitable terms. These would probably include higher wages for blacks and increased taxes, which would raise operating costs for the mines. In addition, seizure of South African interests might be used both for reparations and to provide revenue for a new independent government. Rössing's management has reportedly held talks with SWAPO representatives and apparently believes that it could continue operations under future governments.

The dependence of Namibia's economy on the mining industry suggests that production is likely to continue under any future government, though the continuing potential for violence clearly increases the chances of temporary disruptions of supplies from Namibia.

NIGER

Although the French CEA began exploration for uranium in Niger in the 1950s, the first economically exploitable deposit, Arlit, was not discovered until 1966.[60] At present, two mines are operating in Niger, one other is currently scheduled to begin operation in the next five years, and a number of further deposits are being evaluated. While initial exploration efforts in Niger were dominated by French companies, there has been an active expansion of exploration programs involving numerous countries and companies.

Industry Structure

To date, the only economically feasible deposits are located in the Air Mountains region (see Figure 6-2 for location of mines), and although the Djado Plateau on the Libyan border has shown signs of uranium, the consortium investigating the region has abandoned it due to lack of promising discoveries. The current mining region is a forbidding area, occupied almost entirely by nomads. It was isolated from the rest of the country and the world until the completion of a 600 km all-weather road in 1980. In the past, there have been difficulties in getting trained Nigeriens to work in the remote area.[61] Yet, despite the poor infrastructure, rugged terrain, and great distance from the nearest seaport, the development of the large mines and mills has generally been carried out on schedule. These conditions have, however, made Niger a relatively high-cost producer and thus susceptible to the effects of rapid changes in uranium market conditions.

Arlit. The majority of Niger's production to date has come from the Arlit deposit, discovered by the French in 1966, and operated by the Somair consortium (Société des Mines de l'Air), established in 1968. Initially France's Cogema and other French companies held two-thirds ownership, with Agip and Urangesellshaft each having 8 percent shares and Niger holding a 15 percent interest. In the early 1970s, Somair was a source of dispute when Niger's government, led by Hamani Diori, wished to expand the mine, and the French, because of the poor state of the uranium market, were opposed. This disagreement appears to have spurred Niger's efforts to diversify among exploring companies and customers. In 1975, Colonel Seyni Kountche, by then the ruler of Niger, dissatisfied with the small share of Somair held by Niger, renegotiated the agreement. As a result, Niger's share was boosted to 33 percent, mostly at the expense of Cogema. Current ownership shares are shown in Table 6–5.

Production at Arlit began in July 1971, and capacity was doubled to 1,150 MTU per year in 1975. Between 1975 and 1981, continuous expansion brought capacity to 2,300 MTU. At full production, the deposit would be depleted in the late 1990s. If market conditions warrant, production capacity could be increased to 3,000 MTU annually.

Akouta. The Akouta deposit, discovered in the early 1970s, entered production under the Cominak consortium (Compagnie Minière d'Akouta) in 1978. The mine was constructed by Dowa Mining Co. Ltd. and Nissho Iwai Co. Ltd. for OURD (Overseas Uranium Resource Development, a Japanese consortium of nine utilities established for this operation). Cominak was initially set up as a joint company by Cogema, OURD, and Onarem (Niger's Office National des Ressources Minières), though Cogema subsequently sold part of its share to ENUSA (Spain's Empresa Nacional del Uranio), which receives a share of the output. As with Somair, Kountche's government increased its holding in Cominak after achieving power.[62] Located near the Arlit deposit, Akouta is estimated to contain 44,000 tonnes of uranium, enough to sustain its full production capacity until the end of the century. Capacity was originally planned for 2,000 MTU annually, but was increased to 2,200 MTU by the time production began in August 1978.

Table 6-5. Uranium Deposits in Niger.

Deposit	Partners	Ownership Share (Percent)
Arlit[a]	Onarem	33.00
(Somair)	Cogema	26.96
	CFMU	11.80
	Minatome	7.58
	Mokta	7.58
	Urangesellshaft	6.54
	Agip	6.54
Akouta[a]	Onarem	31.00
(Cominak)	Cogema	34.00
	OURD	25.00
	ENUSA	10.00
Arni[a]	Onarem	50.00
(SMTT)	Cogema	50.00
Azelik[a]	Onarem	50.00
(Abkorun)	IRSA	50.00
Imouraren[a]	Onarem	30.00
	Cogema	35.00
	Conoco	35.00
West Afasto[b]	Onarem	33.30
	Cogema	33.30
	OURD	33.30

Sources:
a. Koutoubi, Sani and Ludwig W. Koch. 1980. "Uranium in Niger." Uranium and Nuclear Energy: 1979. 4th Annual Symposium, Uranium Institute. London: Mining Journal Books.

b. Beaumont, Claude, Director General, Minatome. September 1981. "Uranium in Africa." Presented to Atomic Industrial Forum/Canadian Nuclear Association International Conference on Uranium, Quebec City, September 1981.

Imouraren.   Imouraren is the largest uranium deposit yet discovered in Niger, with reserves estimated at 70,000 MTU.[60,63] Production was originally scheduled to begin in 1984, at a level of 3,000 MTU annually, but the status of the deposit was reassessed by Conoco and Cogema as uranium prices began to fall. Project expenditures were cut to $1 million in 1982 from $6 million in 1981,[64] and Conoco now appears to be seeking a buyer for its share.[65] Cogema also seems to have little interest in increasing supplies from Niger instead seeking to diversify sources outside former French colonies. The low grade of ore and inaccessibility of the deposit have resulted in estimates of production costs at over $35 per pound, well above current uranium prices.

Arni.   The Société Minière de Tassa et de N'Taghalgue (SMTT) has been set up to exploit the small Arni deposit (20,000 MTU reserves). Production has been delayed from 1982–83 to 1985, with full capacity to be reached by 1986.[66] Further delays are likely. Cogema is the operator, sharing ownership on a 50/50 basis with Onarem, and production is foreseen at a rate of 1,500 MTU per year.[67] There are reports that Cogema is seeking a non-French partner, suggesting that it would be difficult for France to absorb its share of uranium production from Arni.[68] Negotiations with Kuwait to invest in SMTT (which was capitalized at $28 million in 1979) have been reported but appear complicated by questions about reimbursement of French exploration expenditures. The state of the market apparently justifies the leisurely pace of such negotiations. It appears likely that Arni will be developed as a follow-on to Arlit, presumably utilizing the Somair mill to reduce new capital investments. This would result in a deferral of startup, possibly until the 1990s, depending on production rates at Arlit.

Azelik.   International Resources, S.A. (IRSA), a subsidiary of the Arabian Oil Co. (operator of the petroleum concession in the Neutral Zone between Kuwait and Saudi Arabia), originally planned to produce 1,500 MTU annually from the Azelik deposit beginning in 1983. Subsequently, operation was delayed to 1986 and planned output reduced to 850 MTU annual production capacity.[66] With the reduction in market prospects, development of Azelik has been postponed until the market improves. Under previous plans, the first 15,000 MTU of reserves were to be strip-mined from the shallow deposit, with a decision being made at a later stage on whether to proceed with deep mining to extend the life of the operation.[69] The output was apparently split equally between the two partners, IRSA and Onarem.

West Afasto.   This deposit, containing about 30,000 MTU of reserves, had been slated for development but has recently been postponed indefinitely, presumably due to market conditions. The operation is owned by Cogema, OURD, and Onarem, in equal shares.

Exploration Activities.   In the late 1970s, after the increase in uranium prices, a number of companies and foreign government entities initiated exploration projects in Niger. As can be seen in Table 6–6, these efforts met with some success. With falling uranium prices and little near-term demand, most of the reserves discovered are unlikely to be developed for some time.

Niger's Trade Patterns

The vast majority of Niger's uranium production has been committed to the equity partners involved in production in the country. However, allocations are not fixed, since Onarem has retained the right to reserve a share of production for independent sales, and other partners have some flexibility in the amounts of uranium they take.

Table 6–6. Principal Exploration Activities in Niger.[a]

Area (Deposit)	Partners	Share (%)	Status
Djado	Onarem	25	Abandoned
	Cogema	25	
	Urangesellschaft	25	
	PNC (Japan)	25	
In-Adrar	Onarem	33	Could come into
	AEOI (Iran)	26	production this
	Cogema	26	decade.
	Agip	15	
East Afasto	Onarem	30	Under evaluation;
(Techili)	BNFL	12	reportedly could
	Nigeria	16	produce this
	Cogema	30	decade.
	Saarberg-Interplan	12	

Source: Koutoubi, Sani and Ludwig W. Koch. 1980. "Uranium in Niger." *Uranium and Nuclear Energy: 1979.* 4th Annual Symposium, Uranium Institute. London: Mining Journal Books.

For Somair at Arlit, Cogema agreed to take 1,000 MTU per year, though this figure was based upon the nominal production level and may vary according to actual production levels. Agip has a ten-year contract for 275 MTU annually, with an option on an additional 275 MTU; however, Agip also has the right to reduce deliveries by 50 to 75 percent each year.[70] Agip did not take its share in 1981, although a special purchase of 600 MTU by the French from Onarem offset the loss. Surplus uranium has been assumed to be Onarem's, although the partners, particularly Cogema, may absorb part or all of it. There is, however, no specific requirement that they do so.

In contrast, the Cominak production agreement appears to include significant conditions on foreign participants, at least in terms of little flexibility in reducing off-take. Under the production agreement, Cogema receives 34 percent of output (equal to its ownership share), 886 MTU annually goes to OURD (greater than its equity share), and 300 MTU is taken by Enusa (also greater than its equity share). These amounts do not appear to have increased when production was increased from 2,000 to 2,200 MTU annually, suggesting an increase in the quantity of uranium that Onarem has to sell.

For the Arni deposit, industry sources indicate that Cogema would receive two-thirds of production (greater than its 50 percent equity share) and Onarem one-third. For Imouraren, it appears that production shares would be the same as equity shares were the venture to proceed. There has been no indication as to whether or not the partners are required to take the surplus from Onarem's share, as with Cominak. The method by which Azelik's uranium would be divided is uncertain.

When spot uranium prices increased dramatically in the second half of the 1970s, the government of Niger appears to have succeeded in increasing the amount of production reserved to it for sales on the spot market. Although initially this proved to be a profitable strategy, the subsequent decline in spot prices has created difficulties for Onarem, complicated by pricing and currency problems. By setting official prices only at the beginning of each year, Onarem has been exposed to lost sales as spot prices have fallen, resulting in a growing disparity between the official price and market prices during the course of each year. Table 6-7 shows the change over time in Niger's official price and the Nuexco Exchange Value. Countering the official price-setting effect has been the depreciation of the French franc (the currency to which Niger's prices are tied), which has led to a reduction in dollar-denominated prices. In fact, had the franc's value not plummeted since 1979, Niger's price would be over $50 per pound of U_3O_8 for 1983. Should the franc recover, it would have a concomitant detrimental effect on Niger's sales unless the official price were reduced significantly.

While falling spot prices make official prices increasingly uncompetitive, the real problem lies with the high cost of producing Niger's uranium. Foreign members of the consortia might choose to see prices kept low, just above variable costs, in order to receive more uranium in return for their sunk-costs, but Onarem has retained a high official price. Thus,

Table 6-7. Niger's Official Uranium Prices *(per lb. U_3O_8)*

	French Francs	U.S. Dollars	Nuexco Exchange Value
1/79	189	45.0	43.25
12/79	189	47.0	40.75
1/80	189	47.0	40.00
12/80	189	41.9	27.00
1/81	154	34.1	25.00
12/81	154	26.8	23.50
1/82	185	32.2	23.00
12/82	185	27.5	20.25
1/83	212	31.5	21.50
12/83	212	26.5	22.50
1/84	225	30.0	20.50

Source: *Nuclear Fuel.* May 24, 1982. "Niger has Raised Its Uranium Price," p. 18; *Quarterly Economic Review of Ivory Coast, Togo, Benin, Niger and Upper Volta.* Second Quarter 1983. p. 23 (Niger); *Nuclear Fuel.* December 5, 1983. "Price of Uranium Produced in Niger Will Rise 6.36%." p. 5.

while the consortia—and the government of Niger—appear to have healthy balance sheets, the foreign participants are paying relatively high prices. The smaller equity buyers have reduced their purchases (where possible) as a result, though for Spain and Italy this was as much due to oversupply as to the price levels. France and Japan appear to have sustained their purchase levels, though probably for different reasons. For France, foreign policy objectives and security predominate. For Japan, Niger's prices are relatively competitive with prices under other long-term contracts, and this source of supply offers diversification and enhanced security. While there has been considerable Japanese interest in reducing procurement costs, the firming of spot prices, which began in late 1982, and the decline of the French franc have helped to maintain the position of Niger's exports in Japan's import slate.

Still, spot-market sales have suffered, and Onarem had difficulty in selling several thousand tons of uranium in the last few years, at least to conventional buyers. Most of Onarem's sales from 1980 to 1982 appear to have arisen in political contexts, compelled by the severe budgetary problems experienced by the nation. France took an additional 600 MTU in 1981 and 800 MTU in 1982, describing it as aid rather than commercial purchases. However, these sales did not entirely reduce Onarem's surplus, and additional sales made by Niger have been controversial.

Late in 1979, the government of Niger began publishing an official journal that included a list of deliveries of uranium and the organization from which they originated (i.e., Somair, Cominak, or Onarem). As a result, the majority of sales made by Onarem since 1979 are known. There is also some information about earlier sales, though this information has not been confirmed officially. Figure 6–5 shows these deliveries for which we have found clear evidence from either official or unofficial sources. Actual deliveries may depart from this schedule. Because we have not been able to account for all of Niger's uranium production prior to 1975, it is likely that spot or other sales (most likely to equity holders) were made. That this material was sold rather than stockpiled is suggested by reports that Niger's 1981 sale of 1,212 MTU to Libya absorbed most of the country's stockpile accumulation. The amount of this sale is very close to the amount of uranium for which we have not been able to account between 1975 and 1981.

Although Onarem's sales have included contracts for small quantities with several European buyers, over 70 percent of the total has been to Libya, most of it in 1980 and 1981. This has caused concern internationally, since Libya has no legitimate use for uranium.[71] Many observers believe much of the uranium bought by Libya prior to 1981 had been shipped to Pakistan (which has refused to accept comprehenisve IAEA safeguards).[72] However, others report a disagreement between the two countries over allegations of Pakistani weapons intentions, leading Libya to withdraw financing. Thus, the final destination of the uranium is not publicly known.

Policy Context

Niger, which achieved independence from France in 1960, is an arid, landlocked republic in west-central Africa. It is the fifth-largest producer of uranium in the free world, with the potential for greatly expanded production. While revenue from uranium exports has been a boon to this impoverished country, its dependence on the uranium mining industry for 50 percent of its gross domestic product and over 90 percent of its export earnings have placed the country's economy at the mercy of the uranium market. Additionally, the strategic nature of uranium means that while Niger can expect protection from France, to whom it will provide a substantial portion of its uranium supply during the 1980s, Niger has also become the object of Libya's potentially aggressive attentions.

Niger gained independence from France in 1960 as a one-party state, the leftist Sawaba (Freedom) Party having been banned by Hamani Diori (the leader of the Niger Progressive Party) in 1959 following his victory in elections (about which questions were subsequently raised).[73] Diori won re-election in both 1965 and 1970, but could not survive the political consequences of a

Figure 6–5.   Niger's Commercial Uranium Exports (*Estimated*).

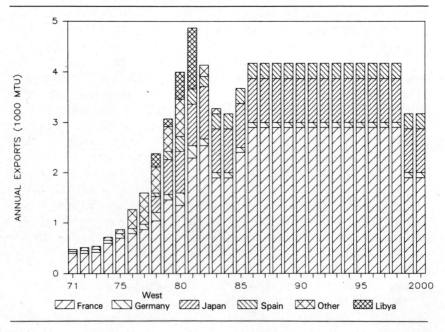

natural disaster. The great Sahelian drought of the early 1970s brought great hardship to a country already 90 percent desert; by 1974, many thousands of people had died, mostly nomads in the northern part of the country. The country's GDP, then almost entirely derived from agriculture, was reduced to less than $50 million, or about $10 per person.[63] The effects of drought were borne mainly by the nomads, who sold or slaughtered their herds in the early stages and subsequently were forced to seek aid from the government. Given the nature of relations between the northern nomads, who historically had raided the south for slaves, and the government (which was dominated by southerners), the Diori government appears not to have been adequately responsive to the problems of the northern nomads.

On April 14, 1974, Lt. Col. Seyni Kountche overthrew Diori's government in a military coup, ostensibly in response to corruption and mismanagement of foreign aid for drought victims. Diori and others were imprisoned and their opposition leadership allowed to return from exile. During the subsequent two years, the Kountche government consolidated its power, putting down two reported coup attempts, one of which involved charges of Libyan assistance to opposition groups. Since 1976, Niger's internal political climate has been relatively stable: in 1980, Kountche freed all remaining political prisoners, including Diori, and announced the formation of a National Commission. The commission was to present a three-year plan for "organizing participation" in the country. The nature of this new political structure is as yet undetermined, though Kountche has suggested a hierarchical system, with villages, urban districts, and nomadic groups at the lowest level, and a national council at the highest. Under this plan, agricultural cooperatives would be established at the village level. Some $100 million was provided in the 1979–83 Five Year Plan to cover the costs of establishing the new administrative system.[73] A national assembly is expected to draft a new constitution in 1983.

While Kountche's preference is for rural cooperatives and a political structure that is neither capitalist nor socialist, he has also made it clear that Niger has no expectation of becoming self-reliant, particularly in the uranium industry. Foreign capital is encouraged, with liberal tax holidays and unrestricted profit repatriation. The primary requirement placed on foreign mining companies is no more than an export tax on the uranium mined. The tax is being used to finance infrastructure; so far, it has gone to building the road from the mining sites to the south. This road is useful both to the mining companies and to establish communications with the interior of the country.[74]

Although data are reported only with considerable delay, it is clear that uranium exports dominate Niger's economy. There is only a small manufacturing sector, largely derived from the processing of agricultural products, and this sector had been repeatedly undermined by drought. Between 1971 and 1976, the value of uranium exports grew from a negligible amount to account for over 60 percent of all exports.[75] This share has subsequently grown to nearly 90 percent, despite a drop in uranium prices.[76] However, government planning had been based on rapidly growing uranium revenues. As a result, there has been retrenchment in recent spending. Infrastructural problems and lack of trained personnel have in the past proven to be more serious constraints than financing, and the originally planned increase in government spending appears to have been unduly optimistic.[77,78] The reduction in planned investment may only bring plans closer to a level that can actually be realized.

The end of the drought of the early 1970s and the development of the uranium industry increased per capita income from $10 in 1974 to $300 in 1980. However, the fall in uranium prices has meant some general economic retrenchment.[79] Since 90 percent of the population is involved in agriculture, a slump in the uranium industry affects the majority of the population only indirectly. However, the impact of this should not be underestimated since much of the government's income goes to irrigation to bring more of the arable land into production. Whether a drop in government funds for agricultural projects would be enough to destabilize Niger politically is not clear. But it is obvious that the government is under tremendous pressure to sell its share of uranium at the best possible price, even if it is to countries like Pakistan and Libya.

Niger's liberal policies toward foreign investment have paid long-term dividends, as prospecting has yielded finds of phosphates, coal, iron ore, and indications of copper and manganese. Small quantities of oil are now being produced near Lake Chad, and a growing mineral sector and energy self-sufficiency are likely in the next few years. Given the undeveloped nature of Niger's economy, even small operations will prove beneficial.

Development of Niger's uranium resources seems unavoidably linked to Niger's external relationships. Libya for some years has had a particularly strong impact on the west African states. One factor believed to have led to the overthrow of Diori was his closeness to Libya, including the signing of a 1974 military agreement and a policy of "arabisation." The 1976 cancellation of most of those agreements by Kountche was followed by a coup attempt possibly backed by Libya, and Kountche has referred to the "insecurity" that has existed between the two nations since then.[80] Periods of rapproachement have alternated with threats from Libya, some revolving around Niger's uranium deposit in the Djado area. Libya issued official maps in 1976, shortly after the coup attempt in Niger, that depicted the Djado plateau as part of Libya, the claim being based on a 1935 border revision.[81] Libya's involvement in the civil war in Chad, which shares a long border with Niger, drove many refugees into Niger and increased fears of politically destabilizing forces.

In addition to Libya's recurrent involvement in Chad (including sending 10,000 soldiers to help one of the factions in 1981, a 1982 attempt to declare a union between Libya and Chad, and a second military involvement in 1983), Libya has actually annexed the Aozou strip along its border with Chad, an area known to contain uranium. Yet, after denouncing Gaddafi and embargoning uranium shipments earlier in 1981, Niger proceeded to sell 1,212 tonnes of uranium to Libya, a country with only a research reactor. Such quantities would be adequate for several small weapons programs, and while it is not obvious that this was an explicit intention, the purchase of a significant amount of uranium brought great international attention to Libya. The 1981 purchase was only made after Kountche expressed concerns about Niger's inability to sell its uranium and just before an Organization of African Unity meeting where Libya was expected to come under strong political attack and needed Niger's support.

The role of France in Niger has been complex. France is the dominant foreign influence, providing much of Niger's foreign aid and access to hard currency, and acting as a major partner in uranium ventures. The French aided Diori in his ascension to power, and he subsequently adopted a strongly pro-France posture, receiving substantial aid in return. However, when French companies resisted expanding the Somair mine, Diori sought to distance himself from the French. Upon his overthrow by Kountche, the uranium agreement was renegotiated to increase Niger's participation in Somair and, despite good relations with France, Kountche successfully sought the participation of other countries in uranium ventures. It seems now essential in French-speaking Africa for leaders not to appear excessively close to the French. Kountche has been careful on this count, although in practice there is continuing reliance on France.

France has been supportive in Niger's conflicts with Libya, promising to aid its ally if threatened. While the nature of this aid has not been established, French assistance seems likely to continue despite governmental changes in France. The Socialist government that came to power in 1981, while insisting that it does not intend to become involved in the internal affairs of its former colonies, is strongly involved in the conflict in Chad and appears likely to sustain an interest in Niger and its uranium.

GABON

Currently the only uranium producer in Gabon is the Compagnie des Mines d'Uranium de Franceville (COMUF), which is jointly owned by the government of Gabon and a consortium of French companies (see Table 6–8). Exploration began shortly after World War II, culminating in the discovery in 1956 of the Mounana deposit. COMUF was established two years later to

Table 6-8.   COMUF Ownership.

Gabon Government	24.75%
Cogema	18.81%
Minatome	13.0%
CFMU (Mokta)	39.98%
Comp. de Generale D'Investissement	3.47%
Gabonese Shareholders	0.99%

Source: Beaumont, Claude, Director General, Minatome. 1981. "Uranium in Africa." Presented to Atomic Industrial Forum/Canadian Nuclear Association International Conference on Uranium, Quebec City, September 1981.

mine this deposit. The Mounana deposit was subsequently exhausted, but neighboring deposits are now being exploited, most notably the Oklo deposit. Table 6-9 lists the estimated reserves of known uranium deposits in Gabon.

Gabon's government maintains a laissez-faire attitude toward COMUF, although it seeks to achieve the greatest possible tax revenue by encouraging a high level of production. In 1976, an agreement was signed providing for a higher tax rate on COMUF and requiring the consortium to build a processing plant to convert the ore fully into yellowcake. In 1978, the plant was completed and, stimulated by high uranium prices at that time, plans were immediately made to raise production capacity to 1,500 MTU annually. By mid-1982, the new capacity was operational, utilizing ore from Boyindzi deposit.

Although uranium production capacity has expanded to 1,500 MTU annually, future production levels are problematical. Since expansion began, no new sales have been made, and the prospects for any in the current market environment are poor. Cogema has been unwilling to absorb additional uranium and so has sought to hold production down. Agip's repeated acceptance of amounts less than the contract maximum has aggravated this

Table 6-9.   Reserves in Gabon.

Deposit	Discovered	Reserves ST U_3O_8	Grade % U_3O_8
Mounana	1956	(5,750)[a]	0.48
Milouloungou	1965	9,940	0.35
Boyindzi	1967	3,270	0.41
Oklo	1968	17,800	0.42
Okelobondo	1974	7,925	0.44

Note:
a. Mined Out.

Source: OECD Nuclear Energy Agency, and the International Atomic Energy Agency. February 1982. *Uranium Resources, Production and Demand.* Paris, France: Organization for Economic Cooperation and Development.

problem. As a result, production seems likely to be held to its former level of about 1,000 MTU annually for the foreseeable future.

Exploration continues to the north of Mounana, in a joint venture between the Korean Electric Company, Cogema, and the government of Gabon, and to the south, by Cogema. Japan's Power and Nuclear Fuel Corporation, along with the government of Gabon and Cogema, has also explored near Libreville. Lower uranium prices and oversupply have greatly reduced incentives to explore further in Gabon. Union Carbide has pulled out of the country altogether, after deciding that its two exploration areas were not particularly promising.[82]

Uranium Trade

Because of the very large role that Cogema plays, it is appropriate to regard Gabon's uranium activities as essentially part of the French system. This was particularly true before 1978, when the yellowcake plant commenced operation, since all preconcentrates were shipped to France for processing. In addition, Gabon's cumulative output through 1982 was only 13,750 MTU, small by comparison with other traditional producers.

Although infrastructural problems make Gabon's uranium expensive compared to new production in Canada and Australia, the French have been willing to purchase from Gabon to sustain supply and continue a long-standing governmental relationship. France has purchased the majority of Gabon's uranium in the past, although it has not entirely absorbed Gabon's surplus, limiting purchases in recent years to about 1,000 MTU annually. Gabon's other customers are mainly European, although some sources indicate Japan has made purchases as well.[83] Italy's Agip signed a ten-year contract in 1977 but has generally taken less than the maximum delivery level specified in the contract. Given Italian uranium surpluses, there are strong incentives for Italy to take reduced quantities in the future or even to cancel the agreement. The withdrawal of Agip would leave Gabon with no other current non-French customers. A summary of known exports and export commitments is given in Figure 6-6.

The decision to reduce production to the level of exports indicates not only that COMUF is unwilling to stockpile uranium in anticipation of a better market, but that it does not anticipate the possibility of sales at favorable prices. Presumably the government is not interested in subsidizing production so that competitive prices can be offered, although past prices have tended to be below those offered by Niger. Reported sales and export figures indicate that the average price was about $25 per pound of U_3O_8 in 1981 and 1982, with substantial fluctuations due to changes in the value of the French franc (i.e., several dollars higher at the beginning of the period and several dollars lower toward the end).

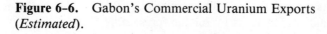

Figure 6-6. Gabon's Commercial Uranium Exports (*Estimated*).

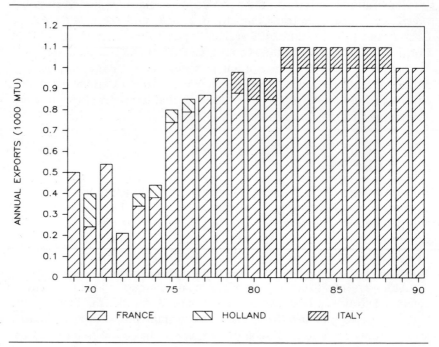

Policy Context

Gabon is a small, relatively wealthy republic of west-central Africa. It formerly was a colonial territory within the federation of French Equatorial Africa. It has only recently become a major exporter of uranium and petroleum and is a member of OPEC.

After achieving independence from France in 1960, Gabon was ruled by Leon M'Ba. M'Ba's rule was consolidated in 1964, when the French intervened under a mutual defense treaty to halt a coup against him by his chief political rival, Jean Hilaire Aubame. New elections were held, and though Aubame's party received a sizeable minority in the National Assembly, it was absorbed into M'Ba's Bloc Democratique Gabonais over the next two years, leaving Gabon as a one-party state since that time.[84]

M'Ba was succeeded on his death in 1967 by his deputy, Bernard-Albert Bongo, who formalized the status of Gabon as a one-party state the following year. Bongo has outlasted several minor domestic threats and territorial disputes with his neighbors, some of whom have accused him of fomenting insurrection against them. He has taken care to maintain political distance from France and began establishing diplomatic relations with the Communist nations in 1974, demonstrating his independence from the West.

Although a loosening of reins in his Parti Democratique Gabonais (PDG) in 1979 led to some criticism of Bongo and the ousting of some high officials, Bongo subsequently received an overwhelming vote of support that December (the result—99.85% in favor—seemed suspiciously high to some). However, opposition candidates from within the party polled 52 percent in a local election when the PDG permitted competition, though only from within the party. Bongo is vulnerable to criticism for filling high government offices with family members, and the gap in national income distribution is quite pronounced. Yet a recent survey by a French organization showed urban residents retained a high level of confidence in him.[85]

Gabon owes its high rate of economic growth largely to its abundant mineral resources (most notably, petroleum), a stable political environment, and policies favorable to foreign investment. Although Gabon found itself badly overcommitted financially after the oil price boom in 1973–74, then heavily in debt when petroleum prices and production subsequently fell, the government has successfully instituted an austerity program that, aided by the International Monetary Fund and the 1979–80 explosion in oil prices, has restabilized the economy.

Gabon's economy is based on the export of natural resources, especially petroleum, timber, manganese, and uranium. Unsuitable land (and tsetse flies) have prevented large-scale, export-based commercial farming.[86] Manufacturing plays a small role in the economy, and is largely confined to the processing of natural resources for export.

Gabon suffers from two serious economic constraints. One is a shortage of skilled labor, with nearly half the population committed to work in subsistence farming. Gabon's desire to prevent ethnic conflicts had led to the expulsion of migrants from neighboring Benin, Equatorial Guinea, and, most recently, Cameroun, thus depriving the country of another source of skilled labor.[87] The second constraint is a lack of infrastructure that has hampered the development of inland resources. Timber in the coastal region has been over-exploited, and reforestation will not yield results for some time. In addition, the development of one of the largest iron-ore deposits in the world, at Mekambo, and the expansion of an extremely large manganese deposit at Moanda, both await the extension of the Transgabonnais railroad, which will not reach these deposits for years. A spur may be sent to Franceville to aid in the export of uranium, which is now sent through the Congo to the south. For uranium, however, low market prices and oversupply are more serious impediments to expansion of production than infrastructure.

Gabon can expect little economic growth in the next few years. But Gabon is already one of the richest African nations in terms of per capita income, and this has helped keep the country politically stable.[88] Thus, although uranium is an important source of revenue for Gabon, it is not as vital as in the other black African producing nations. However, with oil revenue expected to decline due to weak markets, and uranium prices already down, Gabon may face more serious economic problems in the future. Recent discoveries of new petroleum deposits by companies that had been encouraged to resume exploration should at least stabilize production by 1984 and possibly increase it. It is on this, rather than on uranium, that the country's economy will rely for the foreseeable future.

References

1. *Nuclear Fuel.* July 21, 1980. "South Africa Believed Stockpiling U_3O_8 For Possible Fight Against Price-Cutting." p. 4.
2. *Nuclear Fuel.* May 25, 1981. "White Says South Africa Stockpiling is Futile Gesture in Today's Climate." p. 15.
3. For concern about high-grade producers, see: *Nuclear Fuel.* June 22, 1981. "Cutbacks Expected as South Africa Adjusts to Pressure From 'High Grade' Producers." p. 7.
4. For remarks by Oppenheimer, the chairman of Anglo-American, on the industry's future, see: *Nuclear Fuel.* July 21, 1980. "South Africa Believed Stockpiling U_3O_8 For Possible Fight Against Price-Cutting." p. 4.
5. *Mining Journal.* August 1, 1980, Supplement. "Exploration and Expansion." p. 13.
6. For comments on the profitability of uranium by the Minister of Mines, see: *Nuclear Fuel.* January 21, 1980. "Profitability of Uranium in Medium Term Questioned By South African Minister." pp. 14–15.
7. Different sources give different levels. *Nuclear Fuel* (April 12, 1980, and April 18, 1980) reports 1,600 metric tons. NUS Draft Report lists 1,800 MTU_3O_8, and *Mining Journal* (June 29, 1979) puts it at 1,700 MTU_3O_8.
8. *Nuclear Fuel.* November 10, 1980. "Randfontein to Cut Uranium Output by 25% as Processing Shifts From Ore to Tailings." p. 17.
9. *Nuclear Fuel.* November 8, 1982. "Current Uranium Pricing Indicators." p. 15.
10. *Foreign Uranium Supply Update-1980.* Rockville, Md.: NUS Corporation. p. 4–12.
11. *Nuclear Fuel.* July 4, 1983. "East Rand Gold and Uranium Co. Ltd. (ERGO) Plans to Produce 200 Tonnes." p. 9.
12. *Mining Journal.* March 11, 1983. "New Plant for ERGO." p. 157.
13. *Mining Journal.* April 29, 1983, Supplement. "Highlights from the Quarterlies." p. 24.
14. *Mining Journal.* September 14, 1979. "Novel Approaches to Slimes Reclamation at Stilfontein." pp. 209–211.

15. *Nuclear Fuel.* May 11, 1981. "Hartbeestfontein Gold-Uranium Mine Will Suspend Treatment of Accumulated Slimes." p. 13.
16. *Mining Journal.* July 24, 1982. "Uranium: South Africa in Good Health." p. 59; January 28, 1983. "S.A. Golds: Completing the Picture." p. 65.
17. *Mining Journal.* July 30, 1982, Supplement. "St. Helena Gold Mines Limited." p. 17.
18. *Mining Journal.* November 6, 1981, Supplement. "Five Year Review." p. 17.
19. *Nuclear Fuel.* October 16, 1978. "Union Corp's Beisa Mine." p. 4.
20. *Nuclear Fuel.* October 26, 1981. "Analyst Sees Profit in Beisa Uranium At Prices As Low As $12-16 Per Pound." p. 4.
21. *Nuclear Fuel.* September 5, 1977. "Notes on South African Uranium." p. 14; March 7, 1977. "South Africans Report Big Uranium Profits." p. 16.
22. *Mining Journal.* July 10, 1981. "WR Consolidated to Halt Uranium Output." p. 29; October 23, 1981. "South Africa Golds." p. 324.
23. *Mining Annual Review.* 1980. "South Africa/Uranium" Section. London, U.K.: Mining Journal Ltd. p. 496–497.
24. *Nuclear Fuel.* January 7, 1980. "Karoo Area in South Africa Shows Promise." p. 6.
25. *Nuclear Fuel.* March 21, 1977. "Uranium in South Africa: A Report from Johannesburg." pp. 10–11.
26. *Nuclear Fuel.* April 25, 1983. "Japanese Utilities Agree to Pay $32-35/Lb. For Uranium in 1983 and 1984." p. 1.
27. *Mining Journal.* December 19, 1980. "Anglo OFS Mines (Chairman's Review)." p. 509; April 10, 1981. "Golden Opportunities—The Labour Crunch." pp. 269–271.
28. *Mining Journal.* August 7, 1981, Supplement. "Super Mines Galore." p. 17.
29. *Mining Journal.* February 2, 1979, Supplement. "Strong Bullion Demand." p. 11.
30. *Quarterly Economic Review of Southern Africa.* Third Quarter 1981. Namibia Section, pp. 19–28. South Africa Section, pp. 6–19.
31. *Mining Journal.* April 10, 1982. "Golden Opportunities." p. 271.
32. *Quarterly Economic Review of Southern Africa.* Second and Third Quarters 1981.
33. *Africa News.* August 18, 1980. "Namibia's Yellowcake Road." pp. 5–8.
34. *Financial Mail.* Cited in *Africa Research Bulletin.* June 1971. Economic, Financial, and Technical Series. Devon, England: Africa Research Ltd.
35. *South African Financial Gazette.* May 21, 1971. Johannesburg, South Africa: South African Financial Gazette Ltd. Cited in *Africa Research Bulletin.* June 1971. Economic, Financial and Technical Series, Devon, England: Africa Research Ltd.
36. *Star.* May 22, 1971. New York: World News Corporation. Cited in *Africa Research Bulletin.* June 1971.
37. *Financial Times.* June 18, 1976, and September 15, 1976. Cited in *Africa Research Bulletin.* September 1976.
38. *Star.* May 22, 1971. Johannesburg. Cited in Kitazawa, Yoko, "From Tokyo to Johannesburg." February 1975. Unpublished Paper.
39. *Africa Research Bulletin.* June 1980.

210   THE INTERNATIONAL URANIUM MARKET

40. *Foreign Uranium Supply.* 1978. Rockville, Md.: NUS Corporation. pp. 4–12.
41. *Nuclear Fuel.* April 27, 1981. "Rössing Helps RTZ Settle Antitrust Suits for About $22-Million and Still Make Money." p. 11.
42. *Nuclear Fuel.* May 9, 1983. "Production at the Rössing Mine Dropped to 4,910 Tons in 1982." p. 5.
43. *Mining Journal.* December 29, 1978. "Strike at Rössing." p. 507.
44. *Mining Journal.* July 8, 1977. "Gold." p. 25.
45. *Financial Times.* May 12, 1977, and June 24, 1977. Cited in *Africa Research Bulletin.* July 1977.
46. *New African Development.* July 1977. "New Finds in Namibia Bring Uranium Bonanza." pp. 624–625.
47. *Financial Times.* May 20, 1978, and May 26, 1978. Cited in *Africa Research Bulletin.* June 1978.
48. *Nuclear Fuel.* February 1, 1982. "Rössing Uranium Heads for Kansai." p. 4.
49. *Africa Research Bulletin.* November 1975.
50. *Africa News.* August 18, 1980. "Namibia's Yellowcake Road." pp. 5–8.
51. *Nuclear Fuel.* May 28, 1979. "The Rössing Uranium Mine Netted $4-Million." p. 8.
52. *Middle East Economic Survey.* May 4, 1981. Nicosta, Cyprus: Middle East Petroleum and Economic Publications. p. 8.
53. "Namibia (South West Africa): Statistical Survey." 1982. *Africa South of the Sahara.* London, U.K.: Europa Publications Ltd. (12th ed.), p. 728.
54. Rotberg, Robert I. 1980. *Suffer the Future: Policy Choices in Southern Africa.* Cambridge, Mass.: Harvard University Press, p. 183.
55. *Nuclear Fuel.* June 26, 1978. "South Africa-to-France U_3O_8 Airlift Completes Six Months of Operation." pp. 1–2.
56. *Star.* December 19, 1979. Cited in *Africa Research Bulletin.* January 1980.
57. *Quarterly Economic Review of Southern Africa.* Annual Supplement, 1981. London, U.K.: The Economist Intelligence Unit Ltd. p. 36.
58. *Quarterly Economic Review Of Southern Africa.* Annual Supplement, 1981, p. 39.
59. First, Ruth. "Namibia (South West Africa): Economy." 1980–81. *Africa South of the Sahara.* London, U.K.: Europa Publications Ltd. p. 724.
60. *Mining Journal.* September 14, 1979. "Niger Uranium Find Unconfirmed." p. 217.
61. *West Africa.* March 3, 1975. "Kountche in Agades." p. 265.
62. *The Times (London).* Cited in *African Research Bulletin.* October 13, 1973.
63. *West Africa.* January 21, 1980. "What Uranium does for Niger." p. 102.
64. *Nuclear Fuel.* January 18, 1982. "Cogema and Conoco Cut Niger Exploration." p. 11.
65. *Nuclear Fuel.* July 18, 1983. "Conoco Sells Ruby Ranch To Santa Fe Mining and Closes Down Its Minerals Division." p. 4.
66. *Quarterly Economic Review of Ivory Coast, Togo, Benin, Niger, and Upper Volta.* Fourth Quarter 1981. p. 23.
67. *Marches Tropicaux et Mediterranean.* October 5, 1979. Cited in *Africa Research Bulletin.* October 31, 1979.
68. *Nuclear Fuel.* September 29, 1980. "World-Wide French Uranium Efforts Appear to Hedge Against Slowdown in FBR Program." pp. 8–10.

69. *Nuclear Fuel.* May 29, 1978. "Arabian Oil Co. Subsidiary Proposing 1,500 Tonne/yr. U_3O_8 Production in Niger." p. 6.

70. *Nuclear Fuel.* January 3, 1983. "Agip Halves Exploration Budget For Second Time In As Many Years." p. 3.

71. Libya operates only a small research reactor requiring small amounts of highly enriched uranium.

72. *Nuclear Fuel.* December 10, 1979. "Sale of Niger U_3O_8 to Libya, Pakistan Leaves Some Mystery on Intended Use." pp. 3–4.

73. *West Africa.* September 1, 1980. "Niger Moves to End Military Rule." p. 1,654.

74. Most of the foreign aid Niger receives is devoted to either irrigation or road construction, and it has been reported that the Niamey-Air Mountains road was given as "aid" by the Japanese consortium OURD at the suggestion of the government, rather than financed through a tax. *Financial Times.* June 28, 1979. Cited in *Africa Research Bulletin.* October 31, 1978.

75. United Nations. Department of International Economic and Social Affairs. 1979. *Statistical Yearbook 1978/Annuaire Statistique 1978.* New York: United Nations.

76. Author's estimate; official data not available.

77. *West Africa.* July 9, 1979. "Niger's Hopes for Uranium." pp. 210–213.

78. *Quarterly Economic Review of Ivory Coast, Togo, Benin, Niger and Upper Volta.* First Quarter 1981. p. 26.

79. *West Africa.* April 28, 1980. "Niger's Five-Year Plan." p. 731.

80. *West Africa.* December 22/29, 1980. "Libya Provokes Niger to Vigorous Protest." p. 2,610.

81. *West Africa.* October 18, 1976. "Kountche Warns 'Outside Forces'." p. 1,559.

82. *Nuclear Fuel.* January 5, 1981. "Growth of Uranium Sector in Gabon Uncertain in Light of Internal and External Obstacles." p. 10.

83. Robson, Peter. 1981. "Gabon: Economy." *Africa South of the Sahara 1980–81.* p. 393; *Le Soleil.* July 9, 1975.

84. Cornevin, Robert. 1981. "Gabon: Recent History." *Africa South of The Sahara 1980–81.* pp. 389–390.

85. *Quarterly Economic Review of Gabon, Congo, Cameroon, C.A.R. and Chad.* Second Quarter 1981. p. 7.

86. Robson, Peter. 1981. "Gabon: Economy." *Africa South of the Sahara 1980–81.* p. 392.

87. *Quarterly Economic Review of Gabon, Congo, Cameroon, C.A.R. and Chad.* Third Quarter 1981. pp. 7–12.

88. In fact, Gabon is so wealthy that the country became ineligible for developmental aid (which is based on per capita income). This determination was followed by an upward revision of estimates of the country's population by the government and thus a lower estimate of per capita income. However, the United Nations has disputed Gabon's estimates, and a new census is being prepared.

7 SUPPLIER OVERVIEW

In the preceding two chapters we have assessed the industry structure and market roles of the six producer nations playing quantitatively the most important roles in the international uranium market. However, this assessment does not give a complete picture of the supply side of the international market. Not only is there a growing fringe of small producers, but there are two remaining large producers that historically have played only a minor role in the international market but whose actions in the future will have a major effect on that market. The small producers are mostly developing nations (with the exception of Spain and Portugal) whose production will largely be committed to indigenous use. We have already encountered the first of the remaining major producers—France—in our analysis of the roles of Gabon and Niger. France not only imports and consumes large quantities of uranium, but also produces a large amount of uranium domestically and exports to other nations. The second major producer and consumer is, of course, the United States.

In this chapter, we first pull together the assessments of Chapters 5 and 6. We then review briefly the industry and market roles of the smaller producers. Following this, the role of France as a producer and exporter is analyzed. Finally, we consider the relationship between the United States, as a producer and consumer, and the international market. Because the future role of the United States in the international market is so important, we will also return to this topic in later chapters.

MAJOR EXPORTERS

In Chapters 5 and 6, we analyzed the roles of six individual major exporter nations, including industry structure, known exports and export commitments, and potential capacity expansion. In this section, we bring together the information developed in these chapters to present an overview of the dominant suppliers to the international market. Figure 7-1 summarizes the known exports and future export commitments of the six major exporters. The quantities shown may not include spot sales of which there is no public mention, and delivery schedules may be altered somewhat from those shown. However, we do not believe that Figure 7-1 would be significantly altered by such changes.

What is evident from Figure 7-1 is a very rapid growth in commercial exports over the past decade, rising by a factor of thirteen between 1970 and 1983. (It should be noted, however, that there may be early deliveries under general weapons-procurement programs that were used commercially in the late 1960s and early 1970s that are not included in Figure 7-1.) As is evident, significant international commercial trade in uranium has been a relatively recent development. Also evident is a steep decrease in standing export commitments, beginning after 1985.

Many of the commitments shown in Figure 7-1 were made in the mid-1970s, when future supply was highly uncertain, and many of these commitments were for periods of ten years or more. Thus the pattern shown is the reflection of a relatively long-term procurement approach. That the peak in Figure 7-1 is now relatively close in time—not the five to ten years off one would expect if procurement strategies had been maintained—is due to the fact that reactor requirements have been significantly less than the sales shown. As a result, inventories have been building up and, as shown in subsequent chapters, have effectively undermined new sales opportunities for producers.

Since significant commercial exports began in the late 1960s, market shares for the major producers have changed significantly. In Figure 7-2, we show how the part of the international market dominated by the six major producers—the great majority of internationally-traded uranium—has been divided up. In the figure, each nation's exports each year are shown as a percentage of the total for that year. In the early years, Canada accounted for nearly all exports. It was soon joined by Gabon and South Africa. It should be noted here that most of Gabon's exports during this period went to France, where it is difficult to distinguish between use for commercial and military purposes. Thus effective market concentration may have been higher than shown.

In 1970, Niger's exports were added to those of Canada, Gabon, and South Africa. In 1975, Namibian exports began, and in 1976, those from

Figure 7–1. Six Major Uranium Exporters: Estimated Exports and Export Commitments.

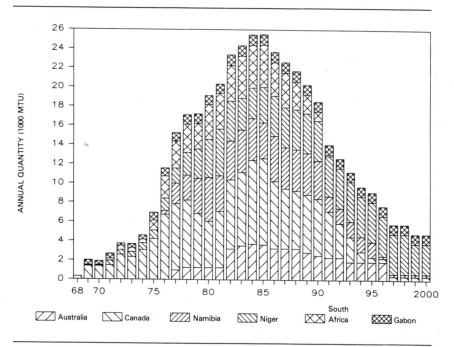

Source: Figures 5–2, 5–6, 6–3, 6–4, 6–5, and 6–6.

Australia. During the years of the cartel (1972–74), Canada had about two-thirds of the market for near-term deliveries, and three other producers controlled the other third in about equal shares, though two of these—Gabon and Niger—were under strong French influence. Since then, the general trend shown in Figure 7–2 is one of greatly increased diversification and reduced market concentration. In the 1980s, Canada still has the largest market share, but other exporters are not far behind. This reduction in concentration goes along with an increase in competitiveness in the international market. It also means that the market has become more resilient to disruptions affecting supply from any single nation.

Under current contracts and equity participations, market shares remain relatively constant for most exporters. The noticeable exception is South Africa, whose share of the market dwindles rapidly in the late 1980s. This reflects a relative lack of long-term consumer supply arrangements with South Africa. As discussed in Chapter 6, South Africa historically has experienced a higher share than most producers of spot or short-term sales.

Figure 7-2.   International Market Shares for Major Exporters (*Estimated Exports and Export Commitments*).

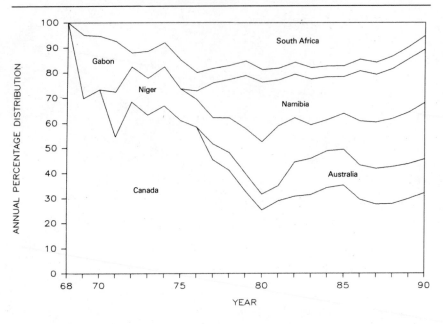

Source: Figure 7-1.

If it continues to do so, revisions of Figure 7-2 made in 1990 will reflect a higher market share for South Africa than that shown. However, while most major exporters prospectively enjoy roughly stable market shares over the decade, South Africa will have to pursue a more aggressive marketing strategy in order to maintain its share. Indeed, inspection of Figure 7-1 suggests that a significant fraction of the drop in scheduled exports after 1985 comes from South Africa.

MINOR PRODUCERS

Geologic indications of uranium have been found in a great many countries. A 1982 review[1] by the OECD Nuclear Energy Agency and the International Atomic Energy Agency assesses prospects in fifty-one nations, and other studies[2,3] consider the resource potential of numerous formations in a host of other countries. However, significant levels of uranium production have occurred, or seem likely to occur, in only a few nations. Uranium is currently being produced in Spain, Portugal, Brazil, and Argentina; production seems likely in the next few years in India and Mexico, though there are likely to be delays for Mexico. Pre-1972 production in Argentina totalled 1,480 MTU; in Spain 170 MTU; and in Argentina 190 MTU. Subsequent production in the six nations considered here is shown in Figure 7-3.

Also shown in Figure 7-3 are the author's estimates of future output. The quantities shown are somewhat conservative, assuming modest efforts at expansion, given surpluses in international markets, and the slowed growth of nuclear power in these countries. In both Portugal and Spain, at least, greater output is feasible. The 1982 OECD-NEA/IAEA assessment saw both countries capable of producing more than 1,000 MTU annually in the 1990s.[1] Actual production seems likely to be determined largely by internal factors in the countries in question—the level of investment possible, domestic nuclear power needs, and, especially for India, the desire for autonomy. In a few cases (such as Portugal, Argentina, and Spain), stocks already exist. These inventories may be available on a loan basis in international transactions, to be repaid in kind when domestic needs increase.

Figure 7-3. Minor Uranium Producers (*Historic Production and Estimated Future Capacity*).

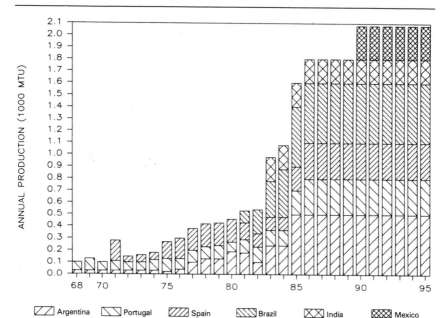

Source: Historic production (Table B-1, Appendix B) and author's estimates.

What is striking about Figure 7–3, however, is the fact that the total quantities shown are barely 10 percent of anticipated actual production for the six major exporters, and only slightly more than 5 percent of production capacity for those six. Given both this disparity, and the likelihood that uranium produced in the smaller states will be used locally, the smaller producers seem unlikely to play a major role in international trade, either in helping determine prices or in offering significant opportunities for diversification of supply for major consumers.

THE FRENCH SYSTEM

As discussed in Chapter 6, France is an important customer of African uranium producers. However, it is also a major producer in its own right, and it exports uranium to Belgium, Sweden, Korea, and other nations. Thus, France, in effect, acts as an important agent in international uranium (and enrichment) trade, importing and re-exporting uranium, primarily from Africa. Figure 7–4 shows this pattern—to the extent that we have been

Figure 7–4. French Imports, Exports and Domestic Production.

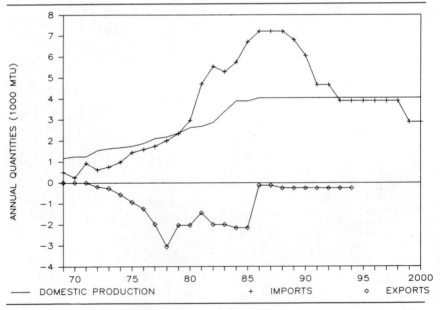

Source: Historic production from Table B–1, Appendix B, and author's estimates of imports and exports.

able to document it—both historically and prospectively. Not including some unknown fraction of the more than 16,000 MTU imported or produced domestically prior to 1969, total French imports through 1982 were about 26,000 MTU, and exports a similarly large 16,000 MTU.

Most of the French export commitments shown were entered into prior to 1974 when tight market conditions and rapidly increasing domestic demand necessitated a freeze on further export commitments. With the return of market surpluses, France has begun to make new commitments for fuel supply, such as that to Korea in connection with French reactor sales. However, it seems unlikely that such export sales will be significant compared to French imports, as was the case in the past. According to the estimates in Figure 7–4, cumulative French imports through 1990 will total nearly 79,000 MTU and exports only about 23,000 MTU.

Thus, while France may continue to play an important role in the international market, it is likely to do so more as a consumer than as a trader. In Chapter 8, we will return to a detailed consideration of French imports, production, and domestic fuel requirements.

UNITED STATES' ROLE

Historically, the United States has had a major effect on the commercial uranium market outside the United States but primarily through its influence on the expectations of other buyers and sellers, rather than through imports or exports. As discussed in Chapter 2, uranium producers that had sold to the United States to meet weapons needs expected to sell to the commercial nuclear power market that began first in the United States. However, the imposition of the U.S. embargo on enrichment of foreign uranium for use in domestic power plants effectively destroyed this expectation, leading to industry maintenance programs in several countries, to reduced investment, and, as we have argued, contributing to the formation of the uranium cartel. The low level of investment in exploration and development outside the United States also helped set the stage for the supply crisis of the 1970s.

The U.S. prohibition on enriching foreign uranium for domestic use was not lifted until 1977 (and then only gradually, as discussed in Chapter 2), prior to which U.S. producers had directly exported 5,300 MTU. U.S. reactor vendors and other buyers had exported comparable (but unpublished) quantities, mostly in connection with reactor export sales. Imports of uranium began in 1975, in expectation of the removal of the ban on enrichment. As shown in Figure 7–5, subsequent imports and exports were quite comparable in magnitude, and both were small compared to domestic

Figure 7-5. U.S. Imports, Exports, and Domestic Supply.

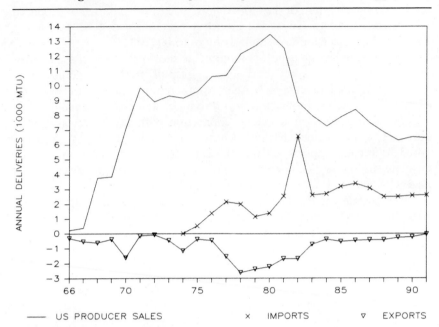

    —— US PRODUCER SALES   x   IMPORTS   ▽   EXPORTS

Source: Historic and scheduled U.S. imports and exports and deliveries from domestic suppliers from: Energy Information Administration. Office of Coal, Nuclear, Electric and Alternate Fuel. September 1983. *1982 Survey of United States Uranium Marketing Activity.* DOE/EIA-0403.

Note: The large import quantity in 1982 arises from a transfer of about 4,000 MTU from Canadian government stocks to Gulf Corporation in the United States in exchange for its properties in Saskatchewan.

deliveries. In 1982, U.S. imports and forward commitments to future imports rose significantly, while U.S. buyers sharply reduced their commitments to domestic suppliers. This shift was in response to changes in basic cost and supply factors, with much of the U.S. uranium industry at a disadvantage in a glutted market dominated by new low-cost production, especially in western Canada. However, the great increase in imports into the United States in 1982 should not necessarily be interpreted as the beginning of a wave of comparable future imports from abroad: the great majority of the uranium imported in 1982 was a large one-time shipment from the Canadian national stockpile by Eldorado Nuclear, in exchange for Gulf Oil Company's share in mining ventures in Canada.

The long-run changes in relationships between U.S. and international markets will be discussed further in Chapter 9.

REFERENCES

1. OECD Nuclear Energy Agency, and the International Atomic Energy Agency. 1982. *Uranium: Resources, Production and Demand*. Paris, France: Organization for Economic Cooperation and Development.

2. OECD Nuclear Energy Agency, and the International Atomic Energy Agency. 1980. *World Uranium: Geology and Resource Potential*. International Uranium Resources Evaluations Project, Phase I Report. San Francisco: Miller Freeman Publications Inc.

3. OECD Nuclear Energy Agency, and the International Atomic Energy Agency. December 1978. *World Uranium Potential: An International Evaluation*. Paris, France: Organization for Economic Cooperation and Development.

PART III
INTERNATIONAL
TRADE: PROBLEMS
AND PROSPECTS

8 SUPPLY TO CONSUMER NATIONS

As of the end of 1982, there were eighteen nations (outside the Centrally Planned Economies or CPEs) operating commercial nuclear power plants with a total capacity of 144 GWe. Of this total, 85 GWe of capacity was outside the United States: 58 GWe in Europe, 16.6 GWe in Japan, and 5.5 GWe in Canada. Only about 5 GWe was operating outside of these industrialized areas. For this reason, most international uranium market activity has been conducted by a few large consumer nations, notably France, West Germany, the United Kingdom, Belgium, Switzerland, and Japan. For many years, developing countries and some other nations with small nuclear programs depended on reactor vendors or fuel fabricators to provide uranium for first cores and initial reloads. Only since 1980 have the developing nations begun to enter the international market directly on their own.

Based on the analyses of Part II (Chapters 5 through 7), it is possible to assemble information about consumer procurements of uranium. To do so, it is necessary to keep track of exports and imports by both the United States and France, as described in Chapter 7. In this chapter, we look first at world uranium procurements (outside the U.S. and the CPEs) by different groups of consumers. Using historic nuclear power growth records and the future nuclear power growth scenarios of Chapter 4, we can then compare these known deliveries and delivery commitments with nuclear plant requirements. In addition, we can estimate requirements for uranium to meet enrichment delivery schedules, often quantities greater than those needed for reactors.

225

These analyses give a global sense of historic and prospective uranium supply and demand balances and allow us to estimate total enriched and natural uranium inventories. However, such a picture does not show in any detail where surpluses and deficits occur; nor does it reveal the variations in procurement approaches undertaken by different groups of consumers. To pursue these questions further, we examine the supply, demand, and inventory situations of the three largest purchasers of uranium internationally: Japan, France, and West Germany. Together these nations account for more than two-thirds of all international commitments to uranium, and nearly 80 percent of those to enrichment.

WORLD SUPPLY AND DEMAND BALANCES

In Figure 8-1, we show known uranium supply agreements to non-U.S. consumer nations with nuclear power programs (non-U.S. WOCA). This tabulation includes any uranium supplied by the United States to other nations, but excludes any uranium supplied by non-U.S. producers to the United States. Thus the total uranium shown is that actually available to non-U.S. consumer nations under known purchase commitments. The reader is reminded that there are likely to be spot sales to consumers that do not appear in the totals. Moreover, there are likely to be delays in some delivery schedules. These two effects work in opposite directions: undiscovered deliveries would add to the quantities shown in Figure 8-1, while delays would push the total delivery curve to the right. Thus the annual quantities shown are probably a reasonably accurate estimation of actual supply until the late 1980s, when deferrals and additional sales may add to the deliveries shown.

Also shown in Figure 8-1 are the supplies committed to various groups of nations. The largest quantities are destined for European consumers, with Japan also enjoying a major share. The Less Developed Country (LDC) group and the "Other" group (including Canadian and South African domestic use, which we include for consistency with later comparisons with reactor requirements) account for only small fractions of the total. Through 1982, the deliveries shown total about 197,000 MTU, 60 percent of which went to Europe,[a] 32 percent to Japan, 4 percent to South African and Canadian domestic uses, and less than 4 percent to the less developed countries. Through 1990, the cumulative total is more than 420,000 MTU, with only a small shift in deliveries toward the LDCs—to a total of 6

[a]This total does not include about 14,500 MTU produced in, or delivered to, France prior to 1969, nor about 1,150 MTU produced in Portugal prior to 1968. Unknown fractions of both quantities were used in the weapons programs of France, the United States, and the United Kingdom.

Figure 8-1.   World Uranium Supply Commitments (*Estimated, Non-U.S. WOCA*).

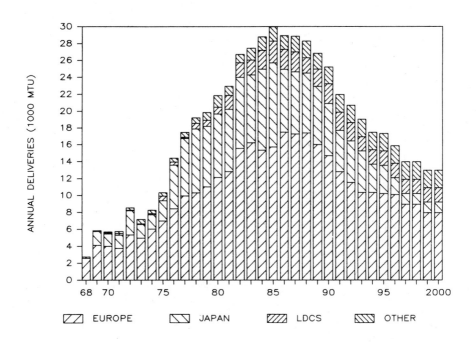

Note: Based on historic and scheduled deliveries identified in Chapters 5 through 7 plus domestic production in France. The Less Developed Country (LDC) group includes Korea, Taiwan, Pakistan, Libya, Iran, Argentina, Brazil and Mexico. "Other" consists of production by South Africa and Canada necessary for domestic use, included here to balance materials flows.

percent. Through the end of the century, there is again a very slight percentage shift toward the LDCs (to 7 percent) on total deliveries of 587,000 MTU.

How does this uranium supply compare with historic consumption and planned and potential nuclear power needs? To answer this question, we have computed reactor requirements for uranium based on historic growth and the two future nuclear power growth scenarios discussed in Chapter 4. As indicated in Chapter 4, the first scenario includes only those reactors to which utilities have made firm commitments. In the next few years, this scenario represents the maximum uranium demand, since it is impossible for utilities to construct more reactors than shown. In the longer term, new orders are possible, as indicated by the Moderate Growth scenario. Figure 8-2 shows

Figure 8-2.   Uranium Supply Commitments Compared to Reactor Requirements (*Non-U.S. WOCA*).

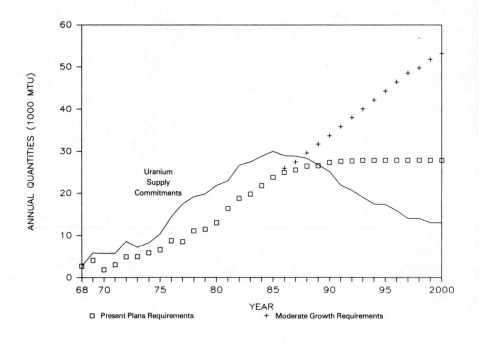

Source: Figure 8-1 and reactor requirements calculation of Chapter 4.

both historic (imputed) reactor requirements and future requirements under the two scenarios, computed according to the assumptions given in Table 1-1. As indicated, annual contracted supply has been significantly above the requirements of existing and committed reactors, and is projected to remain so. Indeed, the excess was about 8,000 MTU in 1982, or as much as 40 percent in excess of annual reactor requirements. Under scheduled deliveries, the excess of annual supply over reactor requirements continues until late in this decade.

This situation will be altered only if changes are made in contracts with producers, or if more reactors are ordered in the near future. Some changes in contracts are already evident. Italy has cancelled some supply arrangements, and Japan and other nations are stretching out deliveries, or taking advantage of contract conditions that may allow reduction (usually of 10 to 15%) in annual deliveries. Many of the changes that have already occurred are taken into account in our summary of current commitments, and thus further opportunities to reduce total world committed supply are somewhat limited without major changes, such as abrogation of contracts.

Perhaps the largest change in supply commitment could come in France, which may reduce its domestic production if it is not possible to reduce imports. In Figure 8-2, we have assumed that French output increases to about 4,000 MTU annually by the mid-1980s; it may be possible for France to limit output to perhaps 2,900 MTU annually—the production rate in 1982. However, this does not remove more than a small fraction of the excess supply. In the next few years, the excess supply might in fact be greater than shown if utilities continue to have difficulty in getting reactors completed and on line, or if reactor performance does not meet the 70 percent capacity factor goal used in our calculations.

It is possible that new utility orders for nuclear power plants could bring supply commitments back into agreement with reactor requirements. Figure 8-2 also shows the uranium requirements implied by the Moderate Growth scenario described in Chapter 4. As shown, the near-term excess supply situation still remains. However, the effect of new reactor orders is to increase reactor requirements to the level of committed supply by about 1987 (rather than 1989 as in the case of no new orders). The new demand in the late 1980s is due to the assumed three-year procurement lead time for first cores for reactors that will begin service in the early 1990s. Thus the date at which annual supply commitments come back into balance with annual reactor requirements will depend critically on new utility orders for power plants over the next few years. If these orders do not come soon, the overall surplus supply situation will continue into the early 1990s.

Even if annual supply and reactor requirements can eventually be brought back into congruence, uranium inventories will still overhang the market well into the next decade. Using the data in Figure 8-2, worldwide stocks at the end of 1982 stood at nearly 75,000 MTU (non-U.S. WOCA) or about four-years forward supply on average. *If* no further reactors are ordered (but utility schedules for ordered plants hold), and *if* consumers take delivery on all the uranium they have committed to, this stock level will peak at about 105,000 MTU in 1989. Under the Present Plans scenario, this would be enough uranium to fuel reactors at that date for well over three years. On average, such inventory levels are not unacceptable. However, as we shall see, inventories are not uniformly distributed among consumer nations.

This prospective stock situation is rather different if new reactor orders are made in the near future, as indicated in our Moderate Growth scenario. In this case, annual reactor uranium requirements begin to exceed contracted uranium supply by 1988—nearly three years earlier than if utilities just build the reactors they have already ordered. Under this scenario, cumulative stocks would be reduced to current levels by 1991 and are used up by some time in 1994.

Figure 8–3 shows the evolution of cumulative worldwide inventories under our two growth scenarios. What is striking about this figure is the dependence of new uranium procurements on new reactor orders. If no new reactors are ordered, and if consumers are willing to reallocate stocks in the secondary market, there will be essentially no need for additional uranium procurements outside the United States until near the end of the century. This situation is quite different if reactor ordering resumes. In this case, there may be a need for as much as 200,000 MTU in additional procurements before the year 2000. Thus, the level of new reactors is a principal determinant of new international demand for uranium.

The Effects of Enrichment Contracting

Current enrichment supply arrangements have an important effect on the uranium market and on the nature of nuclear fuel inventories. Contracts

Figure 8–3.   Cumulative Inventories (*Non-U.S. WOCA*).

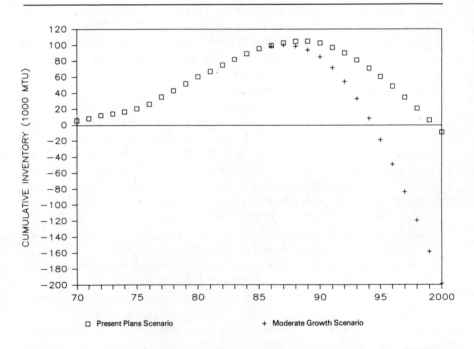

Source: Data of Figure 8–2.

made up to a decade ago require utilities to deliver uranium feed in quantities that are often in excess of reactor requirements, or pay large penalties. These enrichment commitments are partly responsible for the level and pattern of uranium commitments described in the preceding sections, and a significant fraction of excess uranium supply will be held as enriched stock, rather than as natural uranium, unless changes are made in enrichment commitments.

In Figure 8–4, we show total uranium feed requirements associated with the enrichment contracts discussed in Chapters 1 and 4. The calculations assume a 0.20 percent tails assay in enrichment. Also shown are historic and prospective reactor requirements for uranium for reactors using enriched uranium, based on the Present Plans and Moderate Growth scenarios of

Figure 8–4. Enrichment Feed Requirements Compared with Uranium Requirements for Enriched Uranium Reactors and Uranium Deliveries (*Non-U.S. WOCA*).

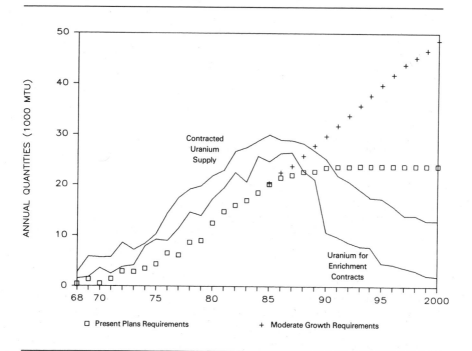

Note: Enrichment feed requirements computed from enrichment commitments of Figure 1–5, assuming 0.20 percent tails assay; uranium deliveries and reactor requirements from Figure 8–2. Uranium feed—totalling about 12,700 MTU—for the 1973 advance sale of enrichment services to Japan is not shown. This material was borrowed from U.S. inventories and subsequently repaid from Japanese commercial uranium purchases.

Chapter 4. Uranium required by reactors not using enriched uranium is *not* included: natural uranium reactors required about 2,700 MTU in 1982, a demand that will grow to about 4,000 MTU annually by the end of the decade. As is evident, uranium feed needed to meet existing enrichment contract and equity obligations is greater than projected reactor requirements for enriched uranium until 1987 or 1988. For example, enrichment feed requirements in 1982 were more than 22,000 MTU, while the uranium needs of reactors requiring enriched uranium were about 17,000 MTU—a difference of more than 5,000 MTU. This difference declines only gradually toward equality over the remainder of the decade.

The delivery of this extra feed has resulted in large stockpiles of enriched material. Assuming 0.20 percent tails assay as the basis for enrichment feed deliveries, the excess feed through 1982 was about 50,000 MTU. Comparing with the total stock number above, we can see that two-thirds of existing inventory is enriched. The enriched stock would fuel existing reactors at the end of 1982 for nearly three years. The investment cost of the 50,000 MTU at current prices (about $65/kilogram U and $130/kilogram SWU) after enrichment would be about $8.2 billion.

As indicated in Figure 8-4, enrichment feed requirements remain greater than reactor enriched uranium needs for some years yet. Inventories of enriched uranium thus continue to grow. For the Present Plans growth scenario, inventories grow until they reach 76,000 MTU (natural uranium equivalent, prior to enrichment) in 1988; for the Moderate Growth scenario, the peak is nearly as high but is reached in 1987. Both amount to more than three years forward supply for reactors operating then. After 1987, enrichment feed imperatives decline rapidly, as shown in Figure 8-4. Existing enriched fuel inventories at that point are sufficient to ensure fuel through 1994 for the Present Plans growth scenario, but only through 1991 for the Moderate Growth case.

In Figure 8-4, we also show total annual commitments to uranium supply (based on Figure 8-2). This supply tracks the enrichment feed requirement curve rather closely until about 1987, and is above that curve only by about the amount of uranium required for natural uranium reactors. This is hardly surprising, given that rational utility procurement practice involved purchases of both uranium and enrichment services. However, it should be noted that uranium commitments were generally made somewhat later than commitments to enrichment, suggesting that enrichment contracting has had some causal relationship to uranium procurement.

Beyond 1988, enrichment feed requirements decline more rapidly than commitments to uranium, falling well below reactor needs by 1990. This represents a change from the high forward commitments of only a few years ago and has been made possible largely by exploiting flexibilities in enrichment contracts with the United States. Thus, by the late 1980s, enrichment contracts cease to be a driving force for uranium procurements. While in the short run, buyers are seeking to defer enrichment commitments because of the surplus shown in Figure 8-4, they will also have to reenter the enrichment market by some time in the early 1990s. If near-term deferrals are not possible, consumers will carry the cost of growing enrichment inventories; otherwise the corresponding deferrals of income will be borne by enrichment suppliers.

Utilities, on average, may not regard stock levels of three- or four-years' supply of enriched uranium as excessive. But there will be pressures to reduce the surplus on the part of some utilities since the surplus is not evenly distributed among consumers. In addition, uranium contracts are also in excess of needs for some consumers; for them, adjustments would have to be made in both uranium and enrichment contracts in order to avoid large enriched or natural uranium stocks. The most likely near-term outcome is an expanded international secondary market in enrichment and enriched uranium. In the longer term, in the 1990s, new enrichment commitments will be necessary. Where and how this will occur is discussed in the next chapter. In the interim, we turn to a discussion of the market roles of key consumer nations.

SUPPLY AND DEMAND SITUATION FOR MAJOR CONSUMER NATIONS

In this section we consider the uranium and enrichment supply situations of the three nations with the largest nuclear programs outside the United States: France, Japan, and West Germany. As noted earlier, these nations account for the great majority of international nuclear fuel trade.

Japan

Japanese utilities have pursued aggressive nuclear fuel procurement policies since the beginning of nuclear power programs. Historically, both uranium and enrichment deliveries have exceeded reactor requirements significantly. Under current supply arrangements, they appear likely to continue to do so for much of this decade.

Japan's first uranium procurements (with the exception of small research purchases from Canada and others) were through the United States in connection with reactor sales by General Electric and Westinghouse. These vendors constructed about 6.5 GWe of capacity in Japan, and imports to Japan from the United States for first cores and initial reloads totalled about 4,000 MTU, virtually all delivered prior to 1980. In 1969,

Japan began independent contracting for uranium, with domestic utilities or groups of utilities seeking uranium abroad.

Imports from Canada and South Africa began in 1969, but significant diversification did not occur until the mid-1970s. In 1974, Japan began to receive uranium from France (through Uranex, the marketing agency of the CEA) on a contract totalling about 9,000 MTU and extending through 1985.[a] In 1977, imports began from Australia and, in 1978, from Niger. Japan also receives substantial amounts of uranium from Rio Tinto Zinc (London), from whom deliveries apparently began in 1977. Some of this uranium might come from RTZ affiliates in Canada and Australia, but RTZ appears to be heavily dependent on Namibia to meet contract obligations to Japan.

Japan thus receives uranium from all major primary producers (with the exception of Gabon) and through Uranex (part of whose supply may come from Gabon). This supply pattern, and its behavior over time, is shown in Figure 8–5. Commitments through the year 2000, including pre-1983 deliveries, total about 154,000 MTU. Of this, about 31 percent comes from Canada, Japan's largest supplier. About 24 percent comes from RTZ, most of which presumably comes from Namibia. Some 9 percent comes from South Africa, 16 percent from Niger, 6 percent from France, 13 percent from Australia, and 4 percent from the United States.

Two trends are evident in Figure 8–5. The first is a successful diversification in supply sources. Whereas in 1973 Japan received 90 percent of its uranium from only two sources (Canada and South Africa), no single primary producer accounted for more than 24 percent in 1982. This diversification clearly enhances Japan's energy supply security. The second evident trend is the declining (relative) importance of supply from South Africa, France, and the United States. While South Africa provided up to one-third of Japan's uranium in the mid-1970s, this share declines to 15 percent by 1980, and to zero by 1990. This shift away from long-term dependence on South African supply may be due to a perception of future insecurity of this supply channel. The end of Japan's dependence on France appears to be due simply to the cessation of French export contracting in 1974. The lack of major new supply from the United States may reflect both disadvantageous prices and a wariness of U.S. export constraints. Given prospective excesses of uranium production capacity worldwide, Japan could contract for additional uranium from virtually all sources.

However, Japan's current supply and demand situation is such that this would not be necessary for quite some time. In Figure 8–6, we show annual

---

[a]These contracts were written prior to the halt of export contracting by France early in 1974. In the initial years, the uranium for Japan would have come from domestic French production or from Gabon or Niger. In the 1980s, some of the contracted uranium could be furnished from additional French supply sources in Namibia, South Africa, or Canada.

Figure 8–5.   Japan's Uranium Imports.

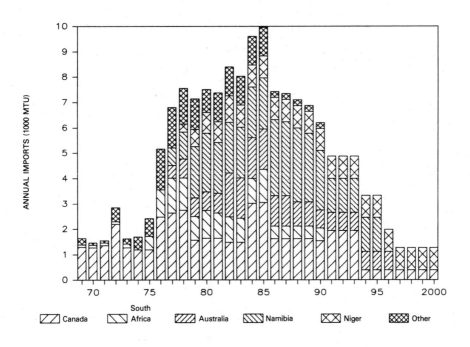

Source: Data of Figures 5-2, 5-6, 6-3, 6-4, 6-5, and 7-4.

consumption requirements in the past, and for our two future growth scenarios. The first scenario is based on plants operating, under construction, and on order—the Present Plans scenario of Chapter 4; the second assumes further reactor orders and is based on the Moderate Growth scenario. Under present utility commitments, only about 25.1 GWe (including a present capacity of about 17 GWe) would be built. The Moderate Growth scenario envisions 37 GWe by 1990 (below what was, until late 1983, the official target of 46 GWe), and 68 GWe by the year 2000. Even these new targets seem optimistic.

Japan's uranium supply commitments have exceeded actual reactor requirements and will continue to do so over the rest of the decade, under both growth scenarios considered here. Our calculations indicate that Japanese reactor consumption of uranium to the end of 1982 has only been on the order of 23,000 MTU, while scheduled delivery commitments have been more than 63,000 MTU (though not all this material has been physically delivered to Japan, and some deliveries may have been postponed). By this

Figure 8-6. Japan: Uranium Requirements Compared to Supply Commitments.

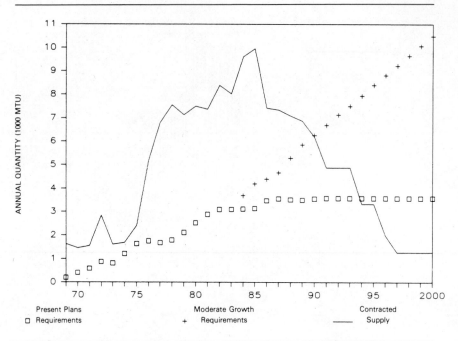

Source: Uranium deliveries from Figure 8-5; uranium requirements calculated from historic nuclear power growth and future growth scenarios of Appendix B, using fuel-cycle assumptions of Table 1-1.

materials balance calculation, current Japanese stocks—including material currently undergoing processing for fuel and that being held for Japan by producers and processors—may be as great as 41,000 MTU (an eight-year supply at 1983 consumption rates).

Based on the Present Plans growth scenario, this stock would grow to about 81,000 MTU by 1993, with contracts exceeding requirements in all years. In contrast, under the Moderate Growth scenario, stocks would peak at about 66,000 MTU in 1989 and decline rapidly thereafter unless new supply arrangements were made. Even so, there would still be stocks left in the year 2000 even if no additional uranium were purchased.

Thus, Japan's uranium supply situation seems secure over the next decade, with existing contracts in excess of the needs of currently planned reactors and, when stocks are taken into account, with enough even for substantial growth in nuclear capacity. Even the loss of a major supplier could be withstood, though this might cause reallocation problems for individual utilities and a disruption of supply logistics. Contract levels and stocks will be sufficient to allow leisure in making new uranium procurement decisions. In addition, the prospective global supply situation is such that there will be many new procurement opportunities. Current Japanese scheduled deliveries decline rapidly in the late 1980s. But the volume then is still greater than would be needed for currently committed reactors. And the stock accumulated by 1990 would be enough to fuel the 25 or so GWe now committed for an additional twenty years. It thus seems likely that Japanese utilities will wait for significant new reactor demand to materialize before committing to new uranium supplies.

The Japanese supply and inventory position is so strong that one must consider whether major readjustments might be made in current supply commitments. Another possibility is that Japanese utilities might enter into sale or loan arrangements much as some U.S. utilities have done. Japanese government policy toward stocks may have a major bearing on utility behavior and on the role of stocks in the world market. Government might encourage large stocks as part of a national energy security program. But given present trends in international uranium markets, security-related concerns have declined in importance, and real (and even nominal) prices have declined greatly. Japan is in a good position to risk some of the higher-priced contracts it now holds by insisting upon downward renegotiation of prices—much as producers insisted upon upward price renegotiations in the tight market of the mid-1970s. In this sense, at least, Japanese stocks overhang the market and may sustain downward pressure on prices.

Enrichment. Enrichment deliveries through 1982 totalled about 28,000 MTSWU. This included 10,020 MTSWU delivered in 1973 under the "advance sale" agreement with the United States. Part of this advanced sale material is still being held for Japan in the United States. By comparison, estimated reactor consumption prior to 1983 was about 13,000 MTSWU, or less than half of total procurements.

The uranium required to meet enrichment contracting requirements through the end of 1982 totalled about 39,400 MTU. This includes about 12,700 MTU feed required for the U.S. advance sale. However, uranium procurements made for delivery through 1982 were even greater, totalling about 63,000 MTU. Since only about 22,000 MTU was used in reactors prior to 1983, it is evident that Japan has built substantial stocks of both enriched and unenriched uranium. While there have been some delays in deliveries, these calculations indicate that about 17,000 MTU may be held as enriched material and 23,000 MTU as unenriched uranium by Japanese utilities or their agents (whether producers, fabricators, or others).

Under current contracts, this procurement pattern continues into the future as shown in Figure 8-7.[a] Enrichment delivery commitments over the

---

[a]Figure 8-7 does not show the 10,020 MTSWU sold to Japan by the United States in 1973.

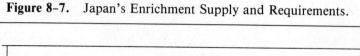

Figure 8-7. Japan's Enrichment Supply and Requirements.

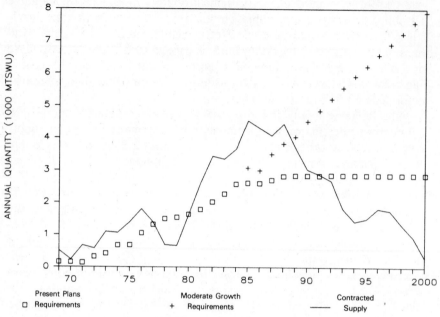

Sources: Enrichment deliveries from Reference 9, Chapter 1; enrichment requirements based on nuclear growth scenarios of Chapter 5 and Appendix B, calculated using fuel-cycle assumptions of Table 1-1.

Note: Does not include 1973 "Advance Sale" of 10,020 MTSWU by the United States to Japan.

period 1983 to 1990 total about 31,000 MTSWU, about 7,000 MTSWU of which will come from Eurodif and the remainder from the United States. This does not include the part of the 1973 advance sale material that will be delivered to Japan after 1982. Under present utility plans (for a total of 25.1 GWe), reactor operations in Japan would require about 22,000 MTSWU during the remainder of the decade, while under the Moderate Growth scenario, the total would be more than 26,000 MTSWU. Thus there will be an excess supply under either growth scenario and enriched inventories will continue to grow.

The uranium feed required to support these future enrichment commitments totals about 39,000 MTU through 1990. However, uranium contracts and firm equity commitments greatly exceed even this enrichment feed requirement, totalling more than 62,000 MTU over the period. Thus, Japan's contracted uranium supply greatly exceeds its enrichment feed requirement, which in turn greatly exceeds prospective reactor requirements. However, while total feed requirements for enrichment may be less than total uranium procurements, the historic pattern of enrichment contracting may have had a strong effect on the schedule of uranium purchases as shown in Figures 8–5 and 8–6. In addition to the large number of Fixed Commitment enrichment sales to Japanese utilities in 1974, the 1973 advance sale of enrichment services required early Japanese uranium purchases. Initially, the U.S. government wanted Japan to buy uranium feed in the United States for the advance sale. Subsequently, the United States agreed to provide the equivalent of 11,700 MTU of natural uranium feed from government stocks, with Japan repaying the uranium loan over subsequent years. Repayments by Japan occurred over the period 1975 through 1981, with the largest quantities in 1976–78 (as much as 3,500 MTU annually). These repayments accounted for a significant fraction of the Japanese uranium purchases shown in Figure 8–5 during this period.

Unless major adjustments are made in both future uranium and enrichment contracting, Japanese stocks of both enriched and unenriched uranium will increase substantially over the remainder of the decade. Under past and current commitments, more than 125,000 MTU will have been delivered to Japan through 1990. Of this, more than 78,000 MTU will have been enriched, and about 47,000 MTU delivered as unenriched material. Of the enriched material, about 48,000 MTU would be consumed under the Present Plans scenario and 59,000 MTU under the Moderate Growth Scenario. Of the 47,000 MTU of unenriched material, less than a thousand tonnes would be consumed in natural uranium reactors.

The Present Plans projection assumes that no additional reactors beyond the 25.1 GWe presently operating, under construction, or on order will be in operation before the first years of the next decade; calculations also assume that no additional commitments are made to enrichment services. Until late 1983, national plans called for 46 GWe of capacity by 1990; this target was revised downward in September 1983 to 37 GWe, a figure identical with our Moderate Growth scenario. Japan is also planning for a full-scale (2 million SWU per year or more) enrichment plant to follow the present pilot centrifuge facility. In addition, if Australia proceeds with an enrichment facility, it may require that uranium destined for Japan be enriched in Australia. But it seems unlikely that any of these nuclear growth or enrichment plans could significantly affect materials balances before 1990. Indeed, it is more likely that official nuclear growth targets will again be revised downward.

There will be increasing economic pressures on Japanese utilities to examine stockpile policies critically. If justification cannot be found for sustaining such large stocks (which might eventually amount to as much as fifteen-years forward supply for the 25.1 GWe of capacity now operating, under construction, or ordered), there may be great pressure for readjustment in contracting. Since uranium commitments greatly exceed enrichment feed requirements, this pressure will probably first be felt in the uranium market, as efforts are made to reduce contract commitments, especially those for high-priced uranium. Near-term changes in enrichment contracting, by this argument, would depend more on national enrichment planning decisions than on stockpile plans, at least until uranium procurement is brought into better congruence with enrichment feed requirements.

Some exceptions to these generalizations may be found among utilities for whom uranium purchases and enrichment feed requirements are in closer agreement or for whom enriched stocks are significantly more difficult to justify (for financial or regulatory reasons) than natural uranium stocks. But enrichment contracting is often not seen in the same light as that for uranium: the latter involve more purely commercial contracts, while the former—especially the enrichment contracts with the United States—involve bilateral political elements and Japanese government participation. Thus pressure for reductions in enrichment commitments will generally be less than for adjustments in uranium commitments for more than one reason.

France

France's role in both uranium and enrichment markets is more complex than Japan's. France is simultaneously an importer, producer, and exporter of uranium; it is also the lead partner in Eurodif, as well as an importer of enrichment services from the United States and the Soviet Union. France also has begun to supply enriched uranium as part of reactor sales to other nations. Finally, France has the largest present commitment to nuclear power outside the United States, making its supply and demand balances quite important to the world market.

Unlike Japan or West Germany, France has substantial domestic production of uranium. France's procurement of uranium abroad is also closely coordinated at the national government level, despite a multiplicity of organizations involved. And the dividing line between military and civilian uranium activities is ambiguous. Analysis of France's position in the uranium market is thus inherently difficult.

Prior to 1969, publicly available data do not allow separation between domestic production and uranium procured from Gabon or other "affiliates"; our estimates indicate that the uranium content of French imports from Gabon from 1961 through 1968 totalled about 3,500 MTU. Total supply through 1968 from all sources was about 16,850 MTU. How much of this was used in the French weapons program is not publicly known. Our calculations indicate that reactor requirements (with appropriate lead times) prior to 1969 totalled about 1,300 MTU, leaving about 15,500 MTU to be accounted for. This quantity should be reduced by actual weapons-related consumption, which may have been on the order of 10,000 MTU. In 1969, it is possible to begin tracking French domestic production and output from Gabon independently. In 1971, production began in Niger, with France receiving most of Niger's output until 1976. Supplies from South Africa began in 1980, and contract deliveries from Namibia and Canada started in 1981. Imports from Australia are scheduled to begin in 1984.

Niger is France's largest external supplier, with imports from Niger reaching more than 2,500 MTU in 1982, and with the potential to grow to about 2,900 MTU annually by 1986 (if the SMTT consortium develops the Arni deposit). France purchased part of the Niger government's share of production in 1981 and 1982, ostensibly for foreign policy reasons. The origins of France's uranium, to the extent they may be deduced from available sources, are shown in Figure 8-8. What is striking about Figure 8-8 is the

Figure 8-8. France: Domestic Uranium Production and Imports.

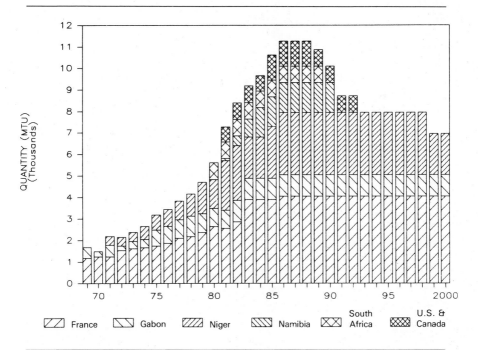

Sources: Data of Figures 5-6, 6-3, 6-4, 6-5, 6-6, and Table B-1, Appendix B.

extent of France's dependence not only on its former African colonies but also on African supply generally. Of total deliveries between 1969 and 1982, all but 8.3 percent came from Niger, Gabon, or France itself, and all but 2.8 percent came from France or Africa. In the early 1980s, France has actively pursued equity positions in non-African supplier nations. French private and state enterprises have invested in ventures in Canada, Australia, and even the United States, as well as pursuing exploration in many developing countries. However, this has not yet altered the long-term supply situation. Of prospective deliveries between 1983 and 2000, only 5.6 percent appear to come from outside France or Africa.

According to Figure 8-8, cumulative French access to domestic and foreign uranium stands at about 70,000 MTU (including some material used for weapons) as of the end of 1982. Annual supply in 1983 will be about 9,000 MTU, and total commitments through the year 2000 may approach 230,000 MTU, depending primarily on France's domestic production levels and imports from Africa. But not all of this uranium is available for domestic use. Between 1969 and 1974, France actively sold uranium in the world market through the CEA-controlled marketing agent, Uranex. Deliveries under commitments made before 1974 (when a hold was placed on new sales) appear to have begun in 1972 and continue until about 1985. There is evidence of about 21,000 MTU of such commitments, largely to Belgium and Japan, though other commitments may exist. Recently France has begun to sell uranium again, in connection with reactor sales, such as those to South Korea. In Chapter 7, Figure 7-4 showed domestic production and known imports and exports. Through 1980, cumulative imports and exports were comparable and French nuclear program needs could be met from domestic production. We may use the data in Figure 8-8 to compute net French supply for domestic uses. Including purchases for weapons, net acquisitions for domestic use through 1982 appear to total about 54,000 MTU and about 131,000 MTU through 1990.

In Figure 8-9, we compare these net supply estimates with historical and prospective reactor requirements. Prior to 1976, net supply significantly exceeded reactor needs. Between 1976 and 1983, net annual procurements were very well matched to annual requirements. However, prospective net supply exceeds reactor needs over much of the remainder of the 1980s. Under the Present Plans growth scenario, French nuclear power capacity growth stops at 52.6 GWe by 1988; under the Moderate Growth scenario, capacity continues to grow through the end of the century. Under both growth scenarios, there is excess supply and stocks continue to grow over the remainder of this decade. However, when allowance is made for military uses, inventories do not appear to grow to more than three- to four-years' forward supply. Thus, unless there are major failures to meet nuclear growth targets, the major problem for France does not appear to be excessive

Figure 8-9.   French Net Uranium Supply Compared to
Reactor Requirements.

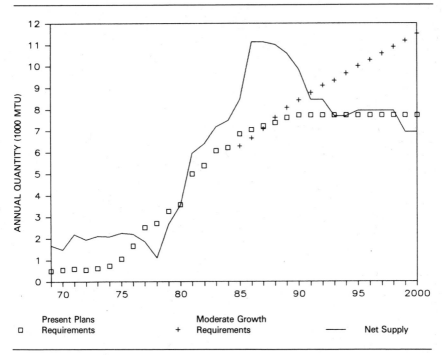

Note: Net supply is domestic production plus imports (shown in Figure 8-8) less exports
(shown in Figure 7-4). Reactor requirements are calculated using fuel-cycle assumptions of
Table 1-1, historic nuclear power expansion, and future growth scenarios of Chapter 4 and
Appendix B.

supply but rather insufficient diversification outside of Africa. What seems
most likely are efforts to replace some African supply with other sources. If
it is difficult to do so for foreign policy reasons, France can simply reduce
domestic production or seek to increase re-exports of uranium to meet its
supply and inventory objectives.

Enrichment. Through 1981, France took delivery of about 19,400
MTSWU of enrichment services, about 23 percent of which came from the
Soviet Union, 12 percent from the United States, and 60 percent from the
Eurodif facility, in which France is the lead partner (small quantities prior
to 1980 were produced by Cogema from other facilities). In the early years
of the French program, the United States was the principal supplier, followed
by the Soviet Union in the mid-1970s; but with the startup of the Eurodif

facility in late 1978, both traditional suppliers became relatively much less important, providing only about 14 percent of France's enrichment supply in 1982. These trends can be seen in Figure 8-10. By comparison, reactor consumption prior to 1983 can be estimated to be about 10,000 MTSWU for domestic French reactors.

Under current contracts, deliveries—at a relatively low level—from the Soviet Union are scheduled to end in 1988 and those from the United States in 1986. At the same time, the enrichment capacity available to France from Eurodif increases rapidly to nearly 5,400 MTSWU annually over the latter half of the decade. Over the remainder of the decade, enrichment delivery commitments to France may be as much as 46,000 MTSWU, about 96 percent of which will come from Eurodif. This assumes that France takes its full share from Eurodif and that the facility operates near its design capacity.

Figure 8-10.   Enrichment Supply and Requirements for France.

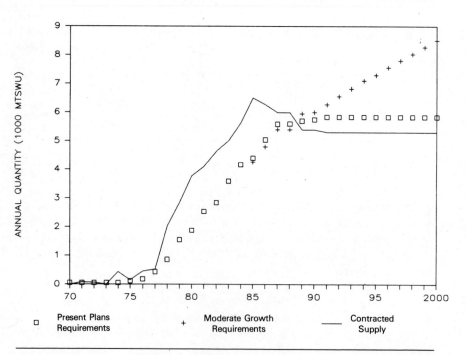

Source: Enrichment deliveries from Reference 9, Chapter 1; enrichment requirements based on nuclear growth scenarios of Chapter 5 and Appendix B, calculated using fuel-cycle assumptions of Table 1-1.

Under the nuclear program envisioned in 1981, total domestic reactor needs over the rest of the decade would total about 42,000 MTSWU. Enrichment export commitments made in connection with reactor sales (to Korea) total about 1,000 MTSWU. Thus, with the nuclear program as earlier envisioned, enrichment supply and demand tend to converge toward the end of the decade and total deliveries would be comparable to reactor requirements. But given the reconsideration of the French program undertaken by the socialist government, there will be delays in the growth of nuclear capacity in France, and thus a reduction in annual delivery requirements. Whether and when France chooses to operate Eurodif at full capacity will depend on stockpile policies, on economic factors associated with investment and variable costs at the plant, and on France's ability to delay or eliminate deliveries from the United States and the Soviet Union. Unless other customers can be found, the cumulative SWU stockpile by the end of the decade would amount to about 14,000 MTSWU. Assuming a variable cost in constant 1982 dollars of about $50 per SWU, the cost of this stock, not including uranium feed, would be about $700 million.

How do uranium delivery commitments compare with enrichment feed requirements and with reactor needs? Of France's net domestic supply through 1982, about 54,000 MTU of natural uranium, an undisclosed portion (probably about 10,000 MTU) was used in France's nuclear weapons program. Total reactor consumption through 1982 can be estimated to be about 31,000 MTU, about 9,000 MTU of which was used in gas-cooled reactors utilizing unenriched uranium. About 31,000 MTU was enriched prior to 1983, only about 22,000 MTU of which was used in enriched uranium reactors. Thus, as of the end of 1982, France's stocks were at most 9,000 MTU held as enriched uranium, and 4,000 MTU held as natural uranium, assuming only 10,000 MTU were used in weapons programs. The allocation of total pre-1983 inputs and outputs of the French system are shown schematically in Figure 8–11. The numbers shown should be regarded as very approximate.

Under the pre-Mitterand nuclear development plan, uranium and enrichment supply and reactor requirements were expected to be in reasonably close congruence. However, with a slowdown in reactor orders, there will be significant annual surpluses of natural uranium and relatively small surpluses of enriched material under current delivery assumptions. Uranium procurement is in some excess, presumably for reasons of security of supply. If desired, France could balance supply and requirements by reducing domestic uranium production. In this way, France could meet its needs from imports when possible, leaving domestic supply stockpiled in the ground. Alternatively, France could reduce imports or increase exports.

Figure 8–11. France: Pre-1983 Cumulative Materials Flow.

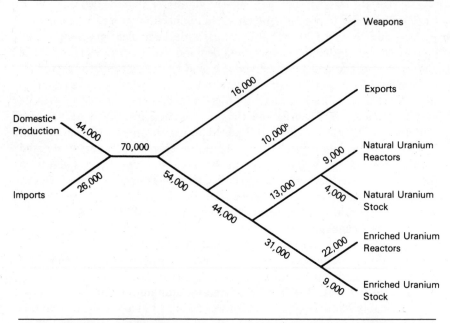

Note: Domestic production figure includes imports from Gabon prior to 1969.
[a]Includes imports from Gabon prior to 1969.
[b]Approximate.

West Germany

West Germany has pursued nuclear fuel procurement practices somewhat different from those of Japan and France. West Germany resembles Japan in having essentially no indigenous supplies of uranium and in having to import enrichment services—at least until Urenco capacity is built up. However, nuclear fuel purchases appear to be in much closer congruence with actual reactor deployment than is the case for Japan.

In contrast to Japan, West Germany did not enter the market early for large quantities of uranium on long-term contracts and, in contrast to France, Germany did not have the opportunity to establish major equity shares in large production ventures until 1980, when it took a share in Australia's Ranger development. Perhaps because of a difference in procurement philosophy, German utilities appear to have contracted only for relatively small quantities in the tight market years of the mid-1970s.

Identifiable deliveries and future uranium supply commitments are shown in Figure 8–12. The total is about 60,000 MTU, nearly half of which was delivered prior to 1983. This might be compared with Japan's commitment to about 154,000 MTU over this same period. In addition, the peak in

Figure 8–12.   West German Uranium Imports.

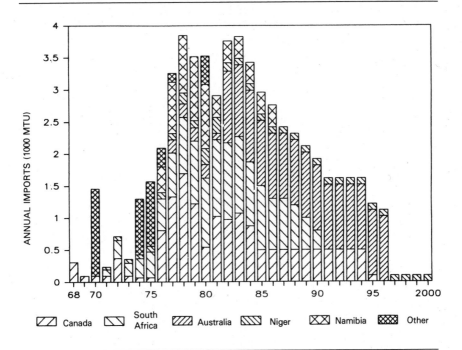

Sources: Data of Figures 5–2, 5–6, 6–3, 6–4, 6–5, and 7–4.

Note: There may be additional spot purchase deliveries to West Germany that are not included here. Other supply came predominantly from the United States in connection with the Offset Agreements discussed in the text.

deliveries for Germany is immediate in time, while that for Japan is not until 1985. Until slowdowns in the German program in the early 1980s, Japanese and German nuclear programs were expected to be very similar in size. Of total German uranium commitments—historic and prospective—more than 26 percent comes from Canada (before 1983, from the Uranerz-Gulf joint venture at Rabbit Lake, and after 1983, from Cluff Lake), nearly 28 percent from Australia, 10.2 percent from Namibia, 5.4 percent from Niger (based on equity shares—additional purchases are possible), 23.5 percent from South Africa, 1.1 percent from France, and 5.8 percent from the United States. Prior to the mid-1970s, Germany's uranium came primarily from Canada, South Africa, and the United States; significant diversification occurred only in the late 1970s.

In Figure 8–13, we compare these known delivery commitments with anticipated reactor needs, under the two growth scenarios described in

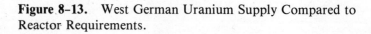

Figure 8–13. West German Uranium Supply Compared to Reactor Requirements.

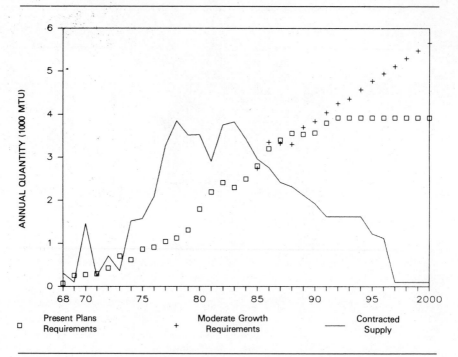

Sources: Uranium deliveries from Figure 8–12; uranium requirements calculated from historic nuclear power expansion and future growth scenarios of Appendix B, using fuel-cycle assumptions of Table 1–1.

Chapter 4. Over the past decade, annual imports have generally exceeded needs, with spot purchases contributing significantly. According to our estimates, German stocks have also grown in each year, albeit slowly, standing at the end of 1982 at about 14,000 MTU. More than 3,000 MTU of this (prior to enrichment) resulted from the "Offset Agreement" of 1968 with the United States. Under this agreement West Germany agreed to purchase enrichment services from the U.S. government and uranium from U.S. producers as a way to balance U.S. expenditures for American troops stationed in West Germany. Two increments have been delivered, and a third (for about 850 MTU) was negotiated in 1980. These known stocks provide a cushion for risks in procurement strategy.

The past approach of German utilities—which was to limit purchases to near-term needs—is now being augmented with efforts to establish more substantial, longer-term positions in the market. Uranerz, Urangesellshaft, and RWE have been active in exploration and joint-development ventures in a number of producer countries. Recent commitments from Cluff Lake and Ranger represent a new emphasis on long-term supply commitments. But under present contracts, stocks still do not appear to grow to the high levels anticipated by Japan, remaining at about a three- to four-year forward supply for German utilities over the entire decade.

The German strategy seems to be to match uranium commitments with actual utility nuclear expansion plans, delaying commitments to uranium until reactors are under construction. Indeed, in the past, reliance was put on spot purchases to provide for at least some of these needs. As long as the international uranium market remains weak, Germany appears likely to continue this approach. The annual needs of reactors in operation, under construction, or on order are covered by annual supply commitments only until 1985, though inventories at that time appear sufficient to meet reactor needs until the early 1990s.

Germany's greatest vulnerability is in its dependence on South Africa and Australia, simply because of the relatively large fractions of supply coming from these two countries. However, this vulnerability does not appear to be a great threat to the German nuclear program. During the Canadian embargo, other deliveries to Germany were adequate to meet reactor requirements without dipping into stocks. And in the future, stocks would be adequate to make up for a loss of any supplier for at least a few years. But the most reassuring fact is that there will be very substantial opportunities to buy more uranium from primary producers and even from other customers.

Enrichment.   Prior to 1983, enrichment deliveries to West Germany totalled about 16,700 MTSWU, about 60 percent from the United States and most of the remainder from the Soviet Union. About 3 percent came from Urenco. These deliveries, including about 3,300 MTSWU purchased from the United States under the 1968 "Offset Agreement" (increments one and three, delivered in 1971 and 1975 respectively; increment two in the early 1980s), intended to balance the dollar flows associated with keeping American troops in Germany.

By comparison, cumulative reactor needs through 1982 were about 8,000 MTSWU, resulting in a significant stockpile, about 25 percent of which (the offset deliveries) is held nationally and the remainder (an average three-year forward supply for present reactors) by utilities. The uranium required to provide feed for these enrichment services was about 23,000 MTU, compared to known uranium purchases by West Germany totalling about 29,000 MTU.

While past nuclear fuel purchases significantly exceeded reactor requirements, future procurements seem intended to stay much closer to anticipated nuclear growth. Total enrichment purchases, under current con-tracts with other suppliers and equity shares in Urenco, for the period 1983–1990 are about 17,000 MTSWU, while reactor needs under present utility plans are for about 19,000 MTSWU. Under enrichment supply ar-rangements as of the end of 1982, annual deliveries from the U.S. DOE and Techsnabexport remain at roughly the same level over the decade, totalling about 28 percent and 29 percent respectively of the total. Historic deliveries and current commitments for future deliveries are shown in Figure 8–14.

The major new source of enrichment supply is Urenco, in which West Germany is a partner with the Netherlands and the United Kingdom. (Small amounts of enrichment are also due from Eurodif from 1983 through 1985, presumably while Urenco capacity is being increased.) The close overall

Figure 8–14. West German Enrichment Supply and Requirements.

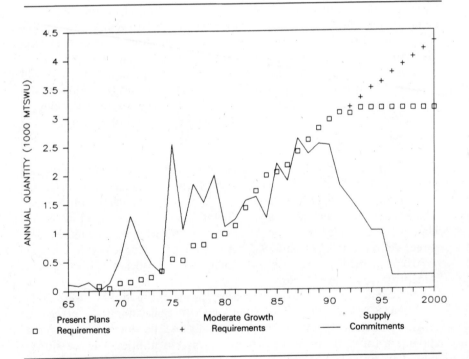

Sources: Enrichment deliveries from Reference 9, Chapter 1; enrichment requirements based on nuclear growth scenarios of Chapter 5 and Appendix B, calculated using fuel-cycle assumptions of Table 1–1.

Note: Large enrichment deliveries shown in 1970s were associated with West Germany's "Offset Agreements" with the United States.

balance between enrichment deliveries and anticipated reactor needs is presumably due to the ability to expand Urenco capacity in small increments: unlike a gaseous diffusion plant, economies of scale are far less important for centrifuge facilities. Unlike France and Japan, West Germany's participation in Urenco allows it greater latitude in adjusting enrichment supplies. However, it should be remembered that West Germany is not entirely protected against the problems resulting from overestimating reactor growth: the expansion of centrifuge capacity has its own momentum, arising largely from commercial commitments to firms manufacturing centrifuges, and a falling away of reactor growth—as seems likely for Germany as for other countries—will mean that real reactor needs may fall below enrichment plant capacity.

Over the remainder of this decade, West Germany appears to have committed to uranium deliveries that are very close to both reactor needs and enrichment feed requirements. Enrichment feed requirements total about 22,000 MTU from 1983 through 1990; forward uranium delivery commitments total about 24,000 MTU (not including possible supply from Key Lake); and total reactor needs over this period are between 24,000 and 25,000 MTU depending on which reactor growth scenario one chooses.

URANIUM INVENTORIES

The general picture that emerges from the preceding analysis is of a general near-term excess of contracted enrichment and uranium supply compared to projected reactor needs. Since inventories are already high, these additional deliveries will result in even larger stocks over the decade, unless major adjustments are made. There is clearly pressure for such adjustment, especially for efforts to delay deliveries of both uranium and enrichment services until later in the decade. Worldwide, current projections of annual reactor requirements begin to exceed firmly contracted enrichment and uranium deliveries by about 1990, though there will still be stocks to be worked off.

However, it appears difficult to make adjustments, at least over the next four or five years. Consumers, in many cases, own equity shares in enrichment plants and even in uranium production ventures. Where enrichment is contracted from independent suppliers, cancellation penalties are large and the political costs for recipient nations appear high.

The readjustment problem varies somewhat from nation to nation. As we have seen, the three major consumers outside the United States have followed rather different procurement strategies. Japan has contracted long term for enrichment services greatly in excess of near-term needs; however, even larger commitments have been made to uranium supply and there is little point in pursuing major changes in enrichment deliveries unless even larger changes can be made in uranium deliveries. Moreover, changing enrichment contracts would have high political costs for Japan.

France has pursued an equally long-term procurement strategy—especially with its investment in Eurodif—and, until recently, has appeared more successful than other nations in coordinating reactor growth with enrichment and uranium deliveries. However, the delays that now seem likely in the French program will tend to create excess supply conditions for France similar to those already being experienced by other nations. West Germany, in contrast, has generally followed a cautious shorter term procurement strategy, with enrichment and uranium supplies closely coordinated with actual nuclear program performance. As a result, West Germany will have smaller adjustment problems.

With a few exceptions, most other nations with nuclear power programs hold enrichment and uranium contracts that exceed possible reactor needs early in the decade, and contracted supply and anticipated reactor requirements become comparable only in about 1990. However, it should be noted that there are great variations among these countries. The predictable result is that much of the activity in both uranium and enrichment markets over the next few years will be in secondary markets, rather than in new contracting activity with primary suppliers. The degree to which adjustment is possible will strongly influence how soon new primary commitments will be made, and where they will be made. But in the long run what will be most important to uranium and enrichment suppliers will be the magnitude of new reactor orders worldwide.

9 THE UNITED STATES IN INTERNATIONAL MARKETS

For many years, there has been only small interchange between the U.S. market for uranium and the market outside the United States. U.S. uranium imports and exports have been roughly in balance and neither has been more than about 10 percent of the trade occurring either domestically or internationally. While the United States still plays a significant role in international enrichment markets, its market share has declined steadily and, given the surplus supply of enrichment worldwide, seems likely to continue to do so. While U.S. utilities have not made significant enrichment purchases directly from foreign suppliers, they have participated in international secondary markets and in arrangements which reduced foreign enrichment commitments from the United States.

As noted in Chapter 7, physical isolation from the world uranium market has not kept the United States from having a major impact on that market, if only through effects on expectations. Following the end of weapons procurements, non-U.S. producers saw the U.S. utility market as the first source of commercial demand for uranium, a hope that was recurrently denied by U.S. prohibitions of imports (for enrichment) for domestic use. And when the United States did enter the world market in 1974, it added to the forces and expectations that subsequently propelled uranium prices upward.

Despite this belated entry, however, actual U.S. import commitments were slow in developing, largely because world prices quickly rose above U.S. marginal costs, removing any competitive advantage for foreign suppliers. In addition, U.S. utilities during this period appear to have avoided foreign supplies, perhaps because of lower perceived supply security or lack of experience in international transactions. Instead, many took equity shares in domestic enterprises or sought other ways to improve access to supplies domestically.

By the early 1980s, these circumstances began to change rapidly. Oversupply and rapidly growing inventories led to a growing secondary market— one-fourth of domestic U.S. deliveries in 1982—and spot prices that rapidly declined. A primary cause of this situation was overexpansion in the domestic U.S. industry. Production in 1980 totaled nearly 17,000 MTU while U.S. reactor requirements (even allowing for considerable lead times) were only about 12,000 MTU; supply and requirements in the United States were badly out of balance. As a result, inventories held by U.S. utilities and producers reached nearly 60,000 MTU by the end of 1982 and annual U.S. production dropped below 10,000 MTU. The U.S. industry entered a depressed state.

While this problem might well have occurred even if there were no uranium industry outside the United States, considerable pressure arose for some restriction on future imports. This pressure was amplified by reduced income, mine layoffs, and other politically visible factors associated with the necessary demand-induced retrenchment in the U.S. industry. However, underlying these more visible signs are two critical factors. The first is the change in resource perspectives described above: with the discovery of low-cost reserves in Canada and Australia, the world supply curve has moved to the right and the present value of U.S. reserves, discounted for the additional years until they become economically attractive, has dropped dramatically. Under open market conditions, many U.S. reserves might now have negligible present value, given the long delays likely until economically feasible exploitation. Severe import restrictions might isolate U.S. producers from this basic economic fact, but at high cost to consumers and with negative impacts on long-run world supply conditions.

The second factor of concern to U.S. producers has to do with the disparate contractual situations of U.S. utilities as compared with most foreign buyers. Many non-U.S. utilities have long-term supply arrangements, involving long-term contracts, equity participation, or other commitments (made either directly or through state enterprises or national lending agencies) that extend five, ten, or more years into the future. In some cases, these arrangements involve front-end financing of mines and mills, as discussed in Chapters 5 and 6.

Such involvements are characteristic of transnational supply arrangements where market and other risks are shared by customers, either through direct investment or through long-term contracts that provide the basis for financing. Because many of these arrangements were made during a period of insecure supply and high demand forecasts, today's aggregate forward delivery commitments exceed reactor requirements even further into the future than might have been originally anticipated. This situation was summarized in Figure 8–2.

In contrast, U.S. utilities have, on average, entered into forward supply commitments that soon drop far below reactor requirements. As shown in Figure 9–1, total annual deliveries from both domestic and foreign suppliers fall quickly below annual reactor requirements. This pattern should be contrasted with the situation outside the United States as shown in Figure 8–2. U.S. utility commitments also seem significantly more flexible than those abroad, with some quantities optional. The forward commitments from

Figure 9–1.   U.S. Domestic Uranium Supply, Import Commitments, and Reported Needs.

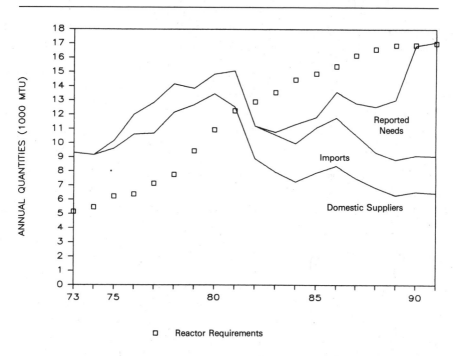

□   Reactor Requirements

Sources: Domestic supply commitments, imports and reported needs from: Energy Information Administration. Office of Coal, Nuclear, Electric, and Alternate Fuel. September 1983. *1982 Survey of United States Uranium Marketing Activity.* DOE/EIA-0403. Reactor requirements are author's calculation based on assumptions of Chapter 1 and the Present Plans scenario of Chapter 4 and Appendix B.

Note: Imports and reported needs are those reported by domestic U.S. utilities. Some reported domestic supply commitments from domestic producers may be met by imports arranged by these domestic suppliers. Import delivery commitments do not include Canadian government inventory transfers to Gulf Corporation in the United States.

domestic suppliers in Figure 9-1 in fact represent a net reduction from those recorded at the end of 1981.

The data shown are from a U.S. Department of Energy survey[1] as of the end of 1982, a year in which deliveries had already fallen below actual reactor requirements. As indicated, imports grow only gradually in absolute terms, though as a percentage of existing supply commitments they increase rapidly. For 1982, DOE figures show an import total greater than that shown in Figure 9-1: an additional 4,270 MTU was imported by U.S. suppliers of uranium, most of which went to Gulf Oil Corporation in exchange for its share in Canadian mine holdings. This material will presumably be used by Gulf to meet existing delivery commitments, some of which are already included in total commitments shown in Figure 9-1 and thus should not be counted twice.

Commitments to imports by U.S. buyers have grown significantly during two periods, as shown in Figure 9-2. The first began in 1974, when it became clear that imports would be allowed beginning in 1976. As discussed above, this new demand on the international market helped produce the increase in world prices. Until 1981, subsequent forward commitments to imports for domestic U.S. use did not increase greatly since higher prices brought forth new domestic supplies. Some material was subsequently resold outside the United States as world prices escalated. However, there were significant increases in forward commitments in both 1981 and 1982. These latest additions to forward commitments are shown in Figure 9-3 (transfers to Gulf in 1982 are not included). As is evident, there were large spot purchases in both 1981 and 1982 but also significant increases in long-term commitments, trebling expected annual deliveries over much of the remainder of the decade. In contrast, U.S. buyer commitments for deliveries from domestic producers (including optional quantities) declined, especially for deliveries in 1982 through 1985, because of cancellations and rescheduling of deliveries to later years.

The difference between the forward time horizons of U.S. and non-U.S. utility contracting indicates that the first portion of the world market that will open to new contracting is the United States. Foreign producers, as was the case in the 1960s, will regard the United States as the only significant near-term market. U.S. producers see the potential for low-cost imports as undermining even their long-term market opportunities, and, again as in the 1960s, wish to have such imports stopped, or at least limited. This issue has been the object of legislative efforts in the U.S. Congress and the subject of several analyses. We will discuss this below.

The DOE survey of U.S. buyers in January 1983 asked these buyers (principally U.S. utilities) to estimate their future purchase requirements. As shown in Figure 9-1, near-term reported needs are relatively small: the sum of total committed supply and reported needs remains below annual reactor

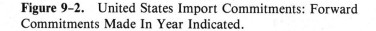

Figure 9-2. United States Import Commitments: Forward Commitments Made In Year Indicated.

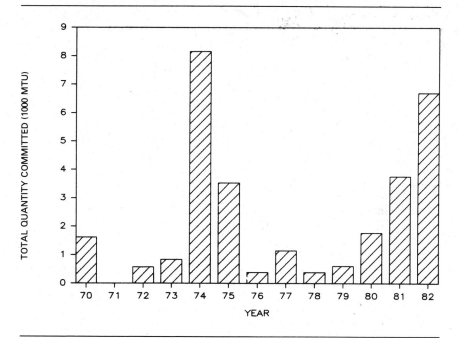

Source: Energy Information Administration. Office of Coal, Nuclear, Electric, and Alternate Fuel. Nuclear and Alternate Fuel Division. September 1983. *1982 Survey of United States Uranium Marketing Activity*. DOE/EIA-0403.

Note: Total for 1981 does not include 7,460 MTU in future deliveries to U.S. buyers resulting from settlements of litigation. Total for 1982 does not include Canadian government stocks scheduled to be delivered to Gulf Corporation in the U.S. in exchange for Gulf's Canadian mine and mill holdings.

requirements until 1990, presumably because utilities plan on working off part of existing inventories before taking delivery on new commitments. Between 1983 and 1991, the reported needs shown in Figure 9-1 total about 29,500 MTU. This figure apparently assumes no reduction in existing delivery commitments. If there are no changes, an assumption examined later, the 29,500 MTU figure may overstate the new market opportunity.

Only those utilities with small inventory positions would report having significant future needs; utilities with very large inventories would not. However, the latter may sell existing inventories to the former in the secondary market. Comparing post-1982 reactor requirements in Figure 9-1 with total supply plus reported needs, we see that utilities apparently plan to work off only about 25,000 MTU between 1982 and 1990. This would leave

Figure 9–3. United States Annual Imports and Import Commitments (*For Domestic Use Only*).

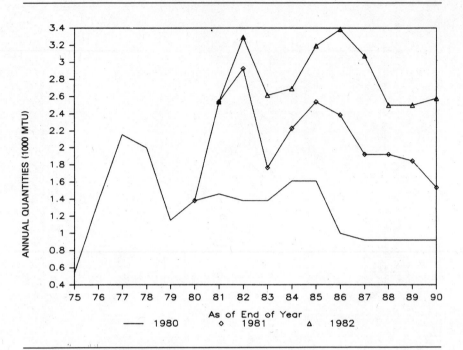

Source: Energy Information Administration. Office of Coal, Nuclear, Electric, and Alternate Fuel. Nuclear and Alternate Fuel Division. September 1983. *1982 Survey of United States Uranium Marketing Activity.* DOE/EIA-0403; U.S. Department of Energy. June 1981 and July 1982. *Survey of United States Uranium Marketing Activity.*

Note: Includes imports resulting from settlements of litigation and those by U.S. uranium companies meeting domestic supply commitments through imports. The 1982 figure does not include the transfer of Canadian government stocks to Gulf Corporation in the United States.

more than 25,000 MTU of additional utility inventory, most likely held by utilities not reporting new needs in the period shown. At least some of this material is likely to be resold to utilities with reported needs, reducing the amount to be demanded of primary producers.

In addition, the secondary market outside the United States seems likely to grow as deliveries to some consumers increase inventory holdings to uncomfortable levels. As noted above, many of these non-U.S. delivery commitments are difficult to change, suggesting that foreign consumers may be at least as eager as primary producers to sell to U.S. buyers. In addition, many of these utilities have contracted presently unwanted enrichment services and may thus be able to offer attractive packages to U.S. buyers. While

U.S. utility commitments to deliver feed to U.S. enrichment plants exceed current uranium delivery commitments plus reported needs (a requirement that could be met from existing inventories, without new purchases), some utilities will be willing to reduce or terminate U.S. enrichment commitments in order to take advantage of discounted uranium and enrichment services from abroad.

In the January 1983 DOE survey, U.S. uranium sellers reported uranium available for sale between 1983 and 1991 at about 30,000 MTU (at prices at or below $60 per pound)—comparable to reported unfilled utility requirements.[2] Since producers reported inventories of about 10,000 MTU, about one-third of the total producer material reported available is already mined. When asked what prices would bring forth what supply, U.S. producers reported very little material available below $30 per pound though a price of $40 would bring more than half of the 30,000 MTU to market (presumably including existing producer inventories). Prices quoted are in year-of-delivery dollars, so there is some uncertainty about what inflation assumptions were used by individual producers. In addition, it is not clear what relationship actual market offerings might bear to survey responses, since producers in a depressed market should be willing to sell at incremental cost (near zero, or even negative due to future carrying costs, for inventories) rather than trying to recover full (sunk) costs.

Contracting and Prices

Contracting between U.S. utilities and domestic suppliers has changed significantly over the past decade, though there are still major differences between U.S. contracting patterns and those abroad. The latter is important to the assessment of the possible future role of U.S. buyers in the international market. Prior to 1974, U.S. domestic uranium procurements were a mix of spot purchases and longer term contracts. The latter generally specified an explicit price at the time of signing, though some allowance was usually made for escalation of production costs and inflation, but not for changes in market conditions. As discussed in Chapter 2, a large fraction of forward commitments in this period were made by Westinghouse, which did not have corresponding supply arrangements with actual producers.

With the great upsurge in prices that began in 1974, a number of these fixed price contracts were renegotiated upward or, especially in the case of Westinghouse, declared void. During the turbulent middle years of the decade, two new types of domestic supply arrangements grew rapidly. One involved direct utility participation in uranium exploration, development, and production. By the end of 1977, thirty of the sixty-five utilities with nuclear power involvements also reported participating in uranium raw material activities, with between 10 and 20 percent of forward delivery commitments (through 1985) under such arrangements.[3] Utility upstream involvement was not just a response to higher prices; it also reflected concerns about supply security and price uncertainty in a period when contracts were unilaterally broken by suppliers.

But even more pervasive was a new emphasis on "market price" contracts. These contracts explicitly took into account not just potential changes in production costs but also the possibility of changes in external market factors. While many such contracts continued to embody a base or floor price (appropriately escalated) set at the time of contract signing, they began to make explicit reference to a market price measured at or near the time of delivery, often the reported Nuexco price or some other measurement of transaction prices contemporaneous with delivery.

At the end of 1976, more than 60 percent of forward deliveries (1977–1985) were still under contracts with prices set at the time the contracts were written; only 18 percent were market price contracts.[4] By the end of 1980, market price contracts had grown to include 36 percent of forward commitments (1980–1990)[5] and, by the end of 1982, to 64 percent of forward commitments (1982–1991). During this period, spot uranium prices had risen, peaked, and then fallen again. The quantities of material delivered and scheduled for delivery (as of January 1, 1983) under various types of contracts are shown in Figure 9–4.[6]

The movement to market price contracts while prices were rising is easy to understand: producers seeing rapidly rising prices wanted to claim the full benefit of such prices, and in a tight market they had the power to obtain such terms. More significantly, many of these contracts imposed rapidly rising floor prices that would continue to go up even if spot market prices failed to do so. In effect, U.S. producers shifted the full burden of market risk to consumers. When spot prices began to fall, it was the escalating floor price that became the basis for setting the "market" price in market price contracts; this floor soon became much higher than open market prices.

For deliveries from domestic producers to domestic buyers during 1982, the average reported price settlement for market price contracts containing a price floor was $51.27 per pound—far above the spot price—while the price settled on for market price contracts without a price floor was only $21.50, very close to the Nuexco spot price. In contrast, the average price for deliveries under traditional contracts, where future prices were set at the time of signing, rose to $38.37.[7] Deliveries from foreign sources during 1982 averaged $25.19 per pound.[8]

According to U.S. survey data, this pattern appears likely to continue into the future. The average price expected (in dollars of the year stated) for traditional contracts rises to about $60 in 1990, with a spread ranging between $45 and $85; the prospective floor price in market price contracts rises

Figure 9-4.   Uranium Deliveries Under Various Contract
Forms.

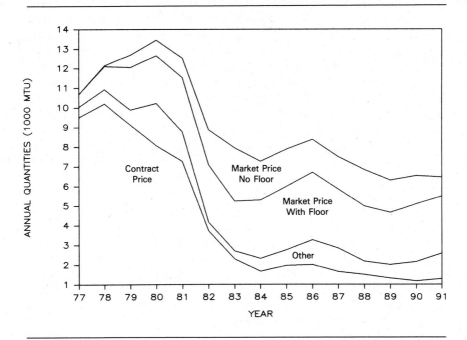

Source: Energy Information Administration. Office of Coal, Nuclear, Electric, and Alternate Fuel. September 1983. *1982 Survey of United States Uranium Marketing Activity.*
DOE/EIA-0403.

to more than $70, with a range from $40 to $120 per pound.[9] The wide
ranges presumably result from different escalation formulas and expectations about input factors, as well as reflecting market conditions and expectations in the year of signing of the contracts. The latter would influence the
base price used for the floor. Based on contract form alone, of course, it is
not possible to project the price in 1990 for deliveries under market price
contracts not containing floor prices.

However, what is clear is that foreign uranium supplies look increasingly
attractive to U.S. utilities faced with rapidly rising contract or floor prices,
given the surplus supply in both the U.S. and international markets. Foreign
purchases may also be attractive to those U.S. suppliers that signed market
price contracts without the price floors that would ensure profitability despite
high (and increasing) domestic production costs. This suggests that not all of
the material shown in Figure 9-4 as domestic procurements will actually come
from U.S. mines.

Of the 74,000 MTU committed for delivery (as of early 1983) from "domestic" sources between 1982 and 1991, about 18,500 MTU is covered by traditional contracts, about 29,800 MTU is under market price contracts with a specified floor, and 17,500 MTU is under market price contracts with no floor (an additional 8,200 MTU is encompassed under captive or other supply arrangements).[10] Of this total, about 11,700 MTU is reported as being optional, and is probably associated primarily with market price contracts. As discussed earlier, U.S. utilities reported unfulfilled needs of about 29,500 MTU over this period. By comparison, import commitments over the same period total about 9,700 MTU. There is clearly substantial room for expansion of imports by both consumers and suppliers in the United States, even at the expense of existing domestic supply arrangements. In the longer term, beyond 1990, the U.S. market is essentially open.

Import Restrictions

Some participants in the U.S. uranium industry have reacted to the potential for increased lower cost imports by asking for government-imposed restrictions on such imports. The political pressure for import limits has been amplified by retrenchment—mine and mill closings, layoffs, and reduced profitability—much of which would have been necessary even without imports, due to overexpansion in the last decade.

The legislative basis for import restrictions was initiated under the 1964 amendments to the Atomic Energy Act of 1954, which provided for the setting of restrictions on imports. These restrictions were to be exercised through limits on enrichment in government facilities of foreign-origin material for domestic use: ". . . the [Atomic Energy] Commission, to the extent necessary to assure the maintenance of a viable domestic uranium industry, shall not offer such services for source or special nuclear materials of foreign origin intended for use in a utilization facility within or under the jurisdiction of the United States."[11] This power was subsequently exercised to exclude imports of uranium for domestic use for more than a decade, with the continuing responsibility transferred successively to the Energy Research and Development Administration and then to the Department of Energy, as these agencies took over many of the functions of the original Atomic Energy Commission.

For purposes of evaluating possible contemporary actions to protect the domestic industry, the 1964 amendments have several deficiencies. One obvious problem is that enrichment is no longer a U.S. monopoly—U.S. utilities might evade restrictions by buying both uranium and enrichment services abroad. This would not only defeat the purpose of the restrictions but also put in even greater jeopardy a U.S. enrichment endeavor already beset by contract cancellations and lack of new demand. A more subtle problem is the lack of specific definitions of "viability." Under a free market system, it would be difficult to argue that the failure of a single firm or group of firms to achieve profitability, or even survival, indicated a lack of economic viability for the industry as a whole; under such a definition, few industries would be "viable." Similarly, a uranium production industry firm might supplement domestic production with imports to meet delivery commitments at lower cost, with imports playing an essential role in maintaining viability and domestic production capability.

A central motive in promoting the viability of the domestic industry has usually been "national security." However, even security is susceptible of conflicting definitions. A threat to national security might be narrowly defined as mere dependence on foreign supplies above some given level. However, it might be defined much more broadly. The chief sources of potential low-cost imports of uranium are Canada and Australia, countries that are important allies and trading partners of the United States. Actions that would endanger relationships with these nations might reduce U.S. security more significantly than any reduction in uranium imports might enhance it.

More generally, import restrictions would add to the pressures that already undermine free trade, perhaps provoking retaliatory restrictions by U.S. trading partners. Finally, isolation of U.S. demand from international supply might adversely affect foreign uranium investment over the next decade, setting the stage for another surge in international prices and undermining fuel supply security in ways that again threaten national and international security.

Several legislative attempts to clarify the conditions under which restrictions might be put on uranium imports eventually resulted in the passage of Public Law 97–415 of January 1983.[12] This legislation specified both a comprehensive presidential review and criteria that should be used by the Department of Energy to evaluate the condition of the domestic mining and milling industry. The comprehensive review included:

1. projections of uranium requirements and inventories of domestic utilities;
2. present and future projected uranium production by the domestic mining and milling industry;
3. the present and future probable penetration of the domestic market by foreign imports;
4. the size of domestic and foreign ore reserves;
5. present and projected domestic uranium exploration expenditures and plans;
6. present and projected employment and capital investment in the uranium industry;

7. an estimate of the level of domestic uranium production necessary to ensure the viable existence of a domestic uranium industry and protection of national security interests;

8. an estimate of the percentage of domestic uranium demand which must be met by domestic uranium production through the year 2000 in order to ensure the level of domestic production estimated to be necessary under subparagraph 7;

9. a projection of domestic uranium production and uranium price levels that will be in effect both under current policy and in the event that foreign import restrictions were enacted by Congress in order to guarantee domestic production at the level estimated to be necessary under subparagraph 7;

10. the anticipated effect of spent nuclear fuel reprocessing on the demand for uranium; and other information relevant to the consideration of restrictions on the importation of source material and special nuclear material from foreign sources.

The secretary of energy was also charged with reporting annually to the Congress and the president for the years 1983 to 1992 "a determination of the viability of the domestic uranium mining and milling industry; and establishing specific criteria to be assessed in these annual reports. The legislation specified criteria to be included:

1. as assessment of whether executed contracts or options for source material or special nuclear material will result in greater than 37.5 percent of actual or projected domestic uranium requirements for any two-consecutive-year period being supplied by source material or special nuclear material from foreign sources;

2. projections of uranium requirements and inventories of domestic utilities for a ten year period;

3. present and probable future use of the domestic market by foreign imports;

4. whether domestic economic reserves can supply all future needs for a future ten year period;

5. present and projected domestic uranium exploration expenditures and plans;

6. present and projected employment and capital investment in the uranium industry;

7. the level of domestic uranium production capacity sufficient to meet projected domestic nuclear power needs for a ten year period; and

8. a projection of domestic uranium production and uranium price levels that will be in effect under various assumptions with respect to imports.

More detailed criteria were developed by the Department of Energy late in 1983, based on hearings held in several regions of the country. Initial suggestions that numerical criteria be set down that would automatically indicate viability were modified to give greater latitude to the secretary of energy in making such a determination. This latitude would, for example, allow consideration of broader security and other interests. However, while the legislation allows the secretary at any time to determine that material "is being imported in such increased quantities as to be a substantial cause of serious injury, or threat thereof, to the U.S. uranium mining and milling industry," he is also mandated, under the 37.5 percent rule, to request the secretary of commerce to initiate an investigation under Section 232 of the Trade Expansion Act of 1962 of the effects on national security of imports of uranium or other source material.[13]

This reliance on the Trade Expansion Act supplements rather than replaces the legal remedies of the viability clause of the Atomic Energy Act. The assessment under the former act takes into account, among other things, the "economic welfare of the essential domestic industry" and the "serious effects of imports on the possible displacement of domestic products, unemployment, decrease in revenues to the government, loss of investments, loss of specialized skills and loss of production capacity." However, the law also demands that the total impact of the proposed action or inaction be taken into account, including "foreign policy considerations, international trade policy, and procurement agreements." It further specifies that the purpose of the investigation is to "safeguard the security of the nation, not the economic welfare of a company or an industry, except as that welfare may affect the national security."[11] Finally, the results of such an investigation are not binding: the president makes the ultimate decision on import restrictions.

At first glance, this elaborate legislative and procedural context does not appear to make import restrictions significantly more likely. Indeed, the sequence of studies and investigations might tend to provide time to resolve issues and inhibit precipitate action on imports. However, it must also be noted that the viability clause of the Atomic Energy Act remains intact and available to those seeking to restrict imports. Under some circumstances, in fact, the determinations required under the new law might provide the basis for invoking the viability clause of the Atomic Energy Act long before investigations under the Trade Expansion Act were completed.

The extent of entry of U.S. buyers—whether utilities, brokers, or uranium companies—into the international market will significantly affect conditions in that market. In the short run—until the end of this decade at least—the United States will be the principal source of new uranium demand for foreign producers and the level of U.S. demand will affect market prices and even market structure. In the longer term, expectations about

U.S. market roles will affect new investment in exploration and production capacity and thus the evolution of the international market. It is to these and other major issues that we now turn.

References

1. Energy Information Administration. Office of Coal, Nuclear, Electric and Alternate Fuel. September 1983. *1982 Survey of United States Uranium Marketing Activity*. Washington, D.C.: U.S. Department of Energy. DOE/EIA–0403. Table 28, p. 45.
2. Energy Information Administration. Office of Coal, Nuclear, Electric and Alternate Fuel. September 1983. *1982 Survey of United States Uranium Marketing Activity*. DOE/EIA–0403, p. 41.
3. U.S. Department of Energy. May 1978. Division of Uranium Resources and Enrichment. *Survey of United States Uranium Marketing Activity*, p. 12.
4. Energy Research and Development Administration. Division of Uranium Resources and Enrichment. May 1977. *Survey of United States Uranium Marketing Activity*, p. 10.
5. U.S. Department of Energy. Office of Enrichment and Assessment. June 1981. *Survey of United States Uranium Marketing Activity*, p. 9.
6. Energy Information Administration. Office of Coal, Nuclear, Electric and Alternate Fuel. September 1983. *1982 Survey of United States Uranium Marketing Activity*. DOE/EIA–0403, p. 19–20.
7. Energy Information Administration. Office of Coal, Nuclear, Electric and Alternate Fuel. September 1983. *1982 Survey of United States Uranium Marketing Activity*. DOE/EIA–0403, pp. 14, 16.
8. Energy Information Administration. Office of Coal, Nuclear, Electric and Alternate Fuel. September 1983. *1982 Survey of United States Uranium Marketing Activity*. DOE/EIA–0403, p. 24.
9. Energy Information Administration. Office of Coal, Nuclear, Electric and Alternate Fuel. September 1983. *1982 Survey of United States Uranium Marketing Activity*. DOE/EIA–0403. Figure 5, p. 17.
10. Energy Information Administration. Office of Coal, Nuclear, Electric and Alternate Fuel. September 1983. *1982 Survey of United States Uranium Marketing Activity*. DOE/EIA–0403. Derived from data in Tables 3, 4, 10 and 11, pp. 8–9, and 19–20.
11. Public Law 88–489, 78 Stat 602. August 26, 1964. An Act to Amend the Atomic Energy Act of 1954. Section 16, p. 606.
12. Public Law 97–415. January 4, 1983. NRC Authorization. 96 Stat 2067.
13. Public Law 87–794. October 11, 1962. Trade Expansion Act of 1962. 76 Stat 872. Section 232 (19 USC 1801).

10 CONCLUSIONS AND CRITICAL ISSUES

The international uranium market has undergone major transformations since its origins in weapons procurement programs following World War II. These procurement programs were highly effective in encouraging new production capacity in the nations that are now the major suppliers to the world market. However, the failure of early nuclear power programs to create comparable commercial demand for uranium—together with what was in effect an embargo of imports by the United States—led to the closing of many mines and overall depression in the international market during the 1960s and early 1970s. During this period, producer governments greatly increased their involvement in uranium industry activities, largely for economic and social welfare reasons.

When panic conditions caused a great upsurge in demand and prices in the middle of the last decade, the international uranium industry again expanded very rapidly, with production doubling in only four years. Where, before 1974, deliveries to consumers had been consistently below not only production capacity but also actual production, supply and demand finally became more closely matched in the last half of the decade. Non-U.S. production and commercial delivery commitments to U.S. and non-U.S. buyers are shown in Figure 10-1. In a sense, this recent period marked the first time that one could speak seriously of a real market, though it was one born under great stresses and distorted by major political interventions.

However, as in the 1960s, the forces that motivated this great industry expansion again led to excess production and the creation of excess capacity. Indeed, capacity outside the United States is still expanding, due largely to

267

Figure 10-1.   Uranium Production and Commercial Deliveries (*Non-U.S. Suppliers to All Buyers*).

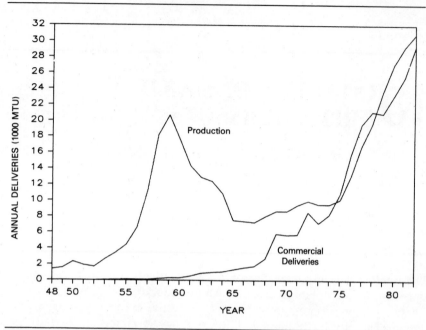

Source: Production from Table B-1, Appendix B. Commercial deliveries are author's estimates based on analysis of Part II of this study.

Note: Excesses of deliveries over production in the period 1974–78 were met from producer inventories.

the development of rich low-cost deposits in Canada. Moreover, during the last decade, consumers took delivery and entered into forward supply commitments greatly in excess of actual reactor needs. As a result, uranium prices fell precipitously after 1979 and secondary market transactions began to displace new procurements from primary producers. In real terms, corrected for inflation, spot prices fell to levels comparable to those of the great depression of the late 1960s and early 1970s, and by 1982 were only half the amount paid for weapons procurements in the mid-1950s. Figure 10-2 shows this price history both in current year dollars, and in constant 1950 dollars for the U.S. market. Prices recorded in the United States in the 1950s in fact do not fully capture the full procurement cost to the government because of other forms of industry support and encouragement; *effective* prices were higher. This only sharpens the contrast. Foreign prices for

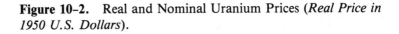

Figure 10-2.   Real and Nominal Uranium Prices (*Real Price in 1950 U.S. Dollars*).

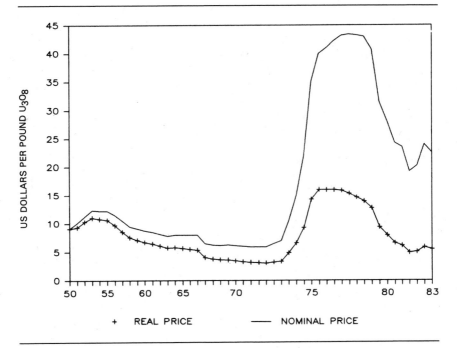

Source: Table 1-4, Chapter 1. Nominal prices in given year dollars, deflated to real 1950 U.S. dollars using U.S. GNP deflator from: *Survey of Current Business.* October 1982 and various 1983 issues. Washington, D.C.: U.S. Department of Commerce, Bureau of Economic Analysis.

weapons procurements were generally higher than in the United States, but subsequently fell to lower levels than shown in Figure 10-2.

During the last decade, uranium reserves also expanded significantly, with the largest changes coming in Australia and in nations not previously considered to hold major low-cost reserves. Figure 10-3 shows these additions to reserves and resources outside the United States (derived from Tables 3-1 and 3-2) since 1967, the first year for which consistent and comprehensive data are available. It also shows annual production outside the United States and changes in spot uranium prices (real price in 1972 dollars). As is evident, rising prices were followed by great increases in known reserves and resources and substantial increases in production. Exploration and development investment also increased with rising prices,

Figure 10–3. Uranium Prices, Production and Additions to Reserves.

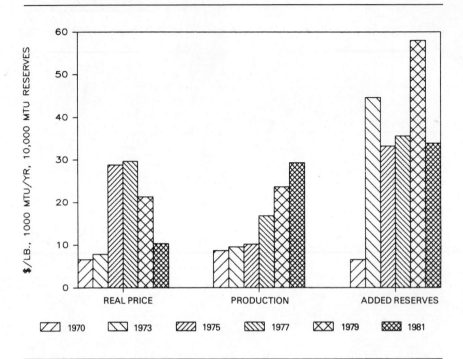

Sources: Prices from Table 4–4, Chapter 4 (real price in 1972 U.S. dollars); non-U.S. production from Table B–1, Appendix B; and additions to non-U.S. reserves from Table B–2, Appendix B.

subsequently falling off after 1979 in most nations. These investments—corrected for inflation—are shown in Figure 10-4. The recent drop in exploration expenditures is hardly surprising given the continuing additions to booked reserves and the even greater decrease in nuclear power demand forecasts.

While increases in known reserves and resources are impressive, the change in resource knowledge that had occurred is even more radical. During the past decade or so, the uranium resource horizon has expanded to include deposits far larger and richer than previously imaginable. By comparison, the sandstone and quartz-pebble conglomerate deposits that provided much of the world's uranium in previous decades now seem exceptionally poor. This has important implications for uranium production in the United States, eastern Canada, South Africa, and other places, where existing reserves should, in effect, be written down or economically depreciated. While con-

Figure 10–4.   World Uranium Exploration Expenditures (*Constant 1972 U.S. Dollars*).

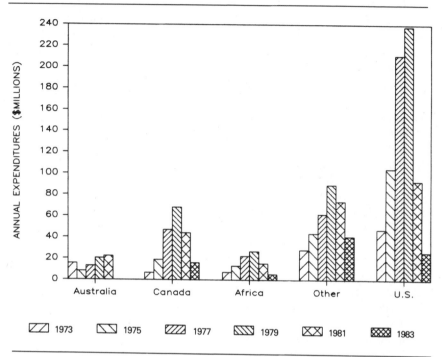

Sources: OECD Nuclear Energy Agency and the International Atomic Energy Agency. 1977, 1979, 1982, 1983. *Uranium Resources, Production and Demand*. Paris, France: Organization for Economic Cooperation and Development; author's estimates. Expenditures for 1983 are IAEA/OECD-NEA estimates.

tinued production from such deposits may be justified in the short run on the basis of variable costs and existing contract prices, new development may not be justified until well into the next century. And even then, massive new lower grade deposits, like that at Roxby Downs in South Australia, may provide economies of scale and opportunities for coproduction of other minerals, whose markets may be counter-cyclical to those of uranium.

The past decade has also seen major changes in international uranium market structure. As shown in Chapter 3, international market concentration was greatly reduced in that more nations more equally share export markets. Industry concentration also declined in most exporter nations, though this came about in part because of increases in government participation in some nations, most notably Canada and Niger, or through imposition of restrictions on foreign equity shares.

In recent years, first in the United States and now internationally, primary producer sales have been supplemented by an active secondary or resale market, involving a host of new actors and old participants acting in new ways. Utilities are reselling uranium (and enrichment and other fuel cycle services) to other utilities and even to producing companies. Some uranium mining companies have found it more profitable to buy material in secondary markets or from foreign producers to meet delivery commitments than to produce uranium from company reserves. In addition, brokers have assumed a larger role in the market through facilitating such transactions. These new developments provide mechanisms to overcome existing market rigidities and increase economic efficiency.

Most of these changes signal a growing maturity in the international uranium market and, perhaps, greater stability of prices and supply to consumer nations. However, the future evolution of this market will remain strongly affected by its historic origins, by the artifacts of its birth, and by many of the same nonmarket forces that historically played such significant roles. The artifacts include the rigidities associated with large investments in uranium production capacity—many of which might not be made today— and long-term contractual obligations to uranium supply and enrichment entered into under assumptions no longer valid. An important issue here is the still unbridged separation between the United States and the international uranium markets. Major nonmarket forces include a significant presence of sovereign intervention, for reasons ranging from nonproliferation to the politics of natural resources.

To expect this world market to function efficiently and without surprises is unrealistic. What is perhaps more useful is to examine more closely the factors that will be most important, and the flexibilities and compensating mechanisms that may alter their impact. Chief among the latter is the existence of known uranium reserves, and production capacity, far in excess of reactor requirements, as well as large inventories and secondary markets that can substitute for primary supply under some circumstances.

In the following sections, we look first at the supply situation, taking into account installed production capacity and potential new additions that might be called upon under appropriate market conditions. Given large uncertainties about such conditions, it is important to know how much flexibility there is in the world supply system. We next look at the problems of near-term adjustment in the market, taking into account the rigidities and imperfections that affect supply and demand over the remainder of this decade, as well as the mechanisms that will ease market adjustment to new circumstances. The central issues over this period concern the extent of sovereign intervention in markets and the nature of U.S. participation in the international market. We conclude with an analysis of longer term market conditions and the outlook for the 1990s.

RESERVES AND SUPPLY POTENTIAL

During the remainder of this decade, and probably through the end of the century, most of the world's uranium is likely to come from mines and mills already constructed, or from deposits already discovered and well understood but which currently lack investment commitments or approvals to proceed. The total potential supply from these sources is significantly greater than likely demand, leaving room for a number of different evolutionary paths for the international market. Which of these paths is most likely, and what conditions might affect the outcome, are the subject of later sections. In this section, we examine the known potential for supply, drawing on the discussion of Chapter 3 and the industry analysis of Chapters 5 through 7. It should be remembered throughout this analysis that not all existing mining ventures can continue, nor all of the possible ventures proceed, under realistic market conditions.

In all traditional producer nations—those discussed in Chapters 5 and 6—as well as in many smaller producing nations, there is already surplus uranium production capacity. However, production capacity is still expanding in some nations, and there is tremendous potential for new output in several countries. Table 10-1 lists existing mines, their current production capacity, potential for expansion, declared reserves, and average ore grade (as the percentage of uranium oxide in ore). The table also lists identified deposits for which there is enough information to establish general characteristics. In most cases, these potential new ventures have been the subject of fairly careful commercial assessment. A number of smaller deposits in major producer nations are not shown, nor are those located in Spain, Argentina, and other nations discussed in Chapter 7. Together, these would add several thousand tonnes to potential annual production capacity.

According to Table 10-1, current total capacity for the nations shown is in excess of 30,000 MTU. To this might be added 3,200 MTU annual capacity in France (expandable to about 4,000 MTU annually) and about 1,000 MTU annual capacity in other nations. Less than 80 percent of this capacity was being utilized at the end of 1983. U.S. production in 1984 seems likely to decline to about half of the peak output of 16,800 MTU in 1980 and the amount of capacity maintained will fall somewhere between these two figures, a function of prospective market conditions.

Under current expansion plans, taking account of deferrals and cancellations through 1983, non-U.S. production capacity would expand to about 42,000 MTU annually by 1990 (not including capacity in Brazil, Argentina, and other nontraditional producing nations). Depletion of existing mines would lead to a modest reduction of capacity during the 1990s. The origins of the capacity expansion are shown in Figure 10-5. In Australia, the exhaustion of Nabarlek is assumed to be compensated for by

Table 10–1. Existing and Potential Mine Developments in Major Producing Nations

	Reserves	Grade[a]	Production Capacity		Cost Category
			Installed	Potential	
Australia					
Ranger	80,000	0.25	2500	5000	L
Nabarlek	10,000	2.10	1500	1500–0	L
Roxby Downs[b]	1,000,000	0.06	—	2500–5000	M–H
Jabiluka	170,000	0.39	—	3800–7600	L
Yeelirrie	42,000	0.14	—	1000–3000	M–H
Canada					
Denison	90,000	0.10	2300	2400	H
Rio Algom	50,000	0.10	2900	4300	H
Rabbit Lake	3,000	0.40	1150	1150–0	L
Collins Bay[c]	20,000	0.38	—	2000	L
Key Lake	70,000	2.70	3000	4600–5400	L
Cluff Lake	16,000	2.45	1500	1000	L
Cigar Lake	70,000	[d]	—	1000–5000	L
Gabon[e]	35,000	0.40	1500	1500	M
Namibia					
Rossing	100,000	0.06	4000	4000	M
Langer Heinrich	40,000	0.12	—	1000+	M
Niger					
Arlit (Somair)	35,000	0.25	2300	3000	M
Akouta (Cominak)	40,000	0.45	2200	2500	M
Imouraren	70,000	0.15	—	3000	H
Arni	20,000		—	1500	M–H
Azelik	15,000		—	850–1500	M–H
South Africa[f]	408,000	0.02	7000	8000	L–M

a. Weighted average percentage of uranium oxide in declared reserves.

b. Uranium at Roxby Downs is to be co-produced with copper and other minerals; uranium production level decisions will have to take this into account.

c. Total for Collins Bay, "A", "B" and "D" ore bodies; weighted average grade suggests wide possible range of output levels.

d. Insufficient data reported; drilling cores shown concentrations up to 25 percent U_3O_8. High ore grade suggests wide possible range of output levels.

e. Gabon total is for Oklo, Boyindzi, Miloulungu and Okelobondo deposits.

f. Total reserves and production capacity for South Africa; see Table 6–1 (Chapter 6) for details. Uranium production level depends in part on economics of coproduction with gold.

Figure 10-5. Minimum Production Capacity Expansion.

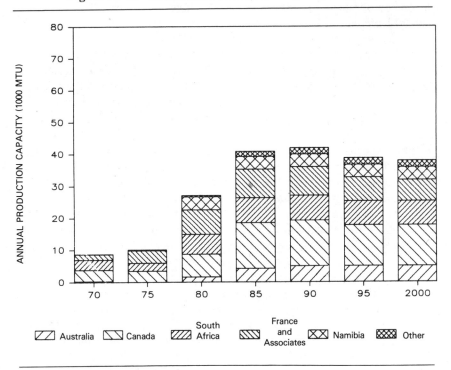

expansion at Ranger or by startup at Roxby Downs, Jabiluka, or Yeelirrie (but only one of these, adding about 2,500 MTU of annual capacity) sometime in 1986.

In Canada, production at Rabbit Lake (where mill throughput has been cut from a nominal 2,000 MTU annual level to stretch out operations) is assumed to end by 1985, with follow-on exploitation of Collins Bay ores deferred until 1991 (when capacity is assumed to be 1,730 MTU). Cluff Lake is assumed to deplete in 1990, Rio Algom is assumed to continue with planned expansion to 4,100 MTU annual capacity by 1985, and Key Lake is assumed to attain an annual capacity of 4,600 MTU by the end of 1984 and add 800 MTU of capacity by 1991. Denison Mines should expand to 2,300 MTU annual capacity and maintain this capacity until the end of the century. As is evident from Figure 10-5, Canada accounts for a significant fraction of the increase in minimum capacity; most of this expansion is seemingly guaranteed by existing and prospective contracts.

For convenience, and because of close associations, capacity in France, Niger, and Gabon is grouped together in Figure 10-5. In Niger, the Somair and Cominak consortia are assumed to maintain existing capacity, but with a drop for Somair to 1,500 MTU annual capacity in 1996 when the Arlit deposit depletes and a switch is made to ore from Arni.

French capacity is assumed to hold at only 2,900 MTU annually, and existing mines in Gabon are assumed to deplete in 1993, with no substitution from other sources. Namibian capacity is assumed to be maintained at current levels, South African capacity is assumed to increase slightly due to completion of recovery facilities being constructed (but with some retrenchment elsewhere), and only modest expansion is projected for minor producers.

This capacity growth scenario makes few demands on the imagination: most of the capacity already exists or is committed under current contracts and the largest additions come from Saskatchewan, where costs and contract prices are low. In a subsequent section we consider how this capacity might be used, or not used. We will also be interested in the potential for more substantial expansion of uranium production capacity, though such expansion obviously would not occur without strong economic incentives.

In Figure 10-6, we have constructed a "maximum production capacity" scenario, with expansion based entirely on known deposits. That is, one need make no assumptions about new exploration or discoveries—such expansion is based on material about which we have relatively good information.

The increases in capacity conjectured in Figure 10-6 come primarily from Canada, Australia, and the French-speaking group of nations. The largest increase, compared to the minimum capacity scenario, comes from Australia, where Roxby Downs is assumed to start up in 1989 and expand annual capacity to 5,000 MTU by 1992. Jabiluka is assumed to expand annual capacity from 2,000 MTU in 1988 to 6,000 MTU in 1995, Yeelirrie to 3,000 MTU by 1992, and Koongarra to 1,000 MTU by 1993. Under these assumptions, annual production capacity in Australia would expand to 20,000 MTU by 1994.

A similar expansion would be possible in Canada. For purposes of this exercise, we assume that Collins Bay ore is used at the Rabbit Lake mill beginning in 1985, that Key Lake expands to 7,400 MTU annual capacity by 1991, that Midwest Lake (or an equivalent venture) begins with a capacity of 1,500 MTU annually by 1990, and that a major new mine (perhaps at Cigar Lake) begins operation in 1993 with a capacity of 1,000 MTU, growing to 5,000 MTU annual capacity by 1996. Even with depletion at some mines, this scenario would maintain a Canadian annual capacity slightly above 20,000 MTU from 1994 onward.

In addition to these increases, it is possible (but probably not attractive) to expand capacity in South Africa (to perhaps 10,000 MTU in the 1990s), to exploit new deposits in Namibia (perhaps with a 2,000 MTU capacity addition at Langer Heinrich or elsewhere) and to sustain current capacity in

Figure 10-6. Maximum Production Capacity Expansion
(*Known Deposits Only*).

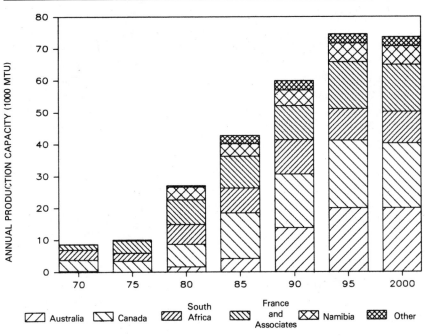

Gabon beyond 1993. In Niger, the Somair consortium might expand existing annual capacity by 3,000 MTU, and Imouraren could go ahead at a
capacity of 3,000 MTU annually by 1993. French production capacity could
be expanded by an additional 1,000 MTU annually and new smaller uranium producers could add a comparable amount of capacity to that assumed
in the minimum capacity growth case.

These production capacity scenarios are rather mechanistic and should
not be taken as predictions. Nor should it be assumed that all this capacity
would be used, even if it came into existence. The most likely world
uranium development path will involve a combination of retrenchment in
some places, with existing capacity disappearing or not being used, and expansion in other areas. How this might occur will be discussed below.

These scenarios are useful, however, in examining two questions. The
first concerns the relationship between supply potential and the level of demand arising under alternative worldwide nuclear power growth assumptions. The second has to do with the impact on the international market of
varying levels of demand from U.S. utilities.

Figure 10–7 shows the situation under assumptions of continued but minimal growth in both uranium production capacity and nuclear power (our Present Plans nuclear growth case of Chapter 4). Through 1982, the graph shows actual production outside the United States as well as non-U.S. consumer purchases. The second line indicates the additional deliveries made to the United States—reported historic imports for domestic use. Beyond 1982, the upper line is the total capacity envisioned by the minimum uranium production capacity scenario described above. To estimate uranium demand beyond 1982 for non-U.S. utilities, we have used actual scheduled contractual deliveries through 1989. During this period, scheduled deliveries exceed reactor requirements, so that consumer inventories continue to build. Beginning in 1989, the curve shows actual non-U.S. reactor requirements.

Figure 10–7.   Demand On Non-U.S. Producers for Planned Reactors (*Minimum Uranium Production Capacity*).

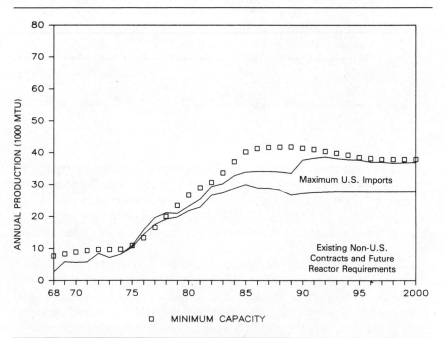

Note: Assumes minimum capacity expansion of Figure 10–5. Demand is assumed equal to existing contracts through 1989 for non-U.S. utilities and thereafter equal to actual reactor requirements under the Present Plans scenario. Under these assumptions, non-U.S. utilities continue to build inventories until 1990 and hold these inventories beyond that date. U.S. utilities are assumed to take delivery under existing import and domestic contracts and to meet reported "uncovered needs" through additional imports until 1992. After that date, U.S. utilities are assumed to meet half of their needs through imports.

For the United States, import demand through 1991 is assumed to consist of two parts: actual scheduled import deliveries, as reported to the Department of Energy, and similarly reported "uncovered needs." That is, U.S. buyers, for purposes of this illustration, are assumed not to buy any additional supplies domestically through 1991. After that date, the figure assumes that half of total U.S. reactor requirements are met by imports.

These assumptions probably overstate total forward international demand on primary uranium suppliers—consumers may back out of some contracts or reduce deliveries, utilizing their own inventories or purchasing those of others in the secondary market. While it is possible that U.S. utilities will escape domestic delivery commitments and go abroad for replacement supplies, the assumptions made here most likely overstate U.S. demand on the international market, at least until the 1990s. What is most interesting about this construction is that even optimistic demand assumptions do not bring the international uranium industry to full capacity utilization until near the end of the century, unless new reactors are ordered.

New reactor orders would obviously increase future demand for uranium. To explore this effect, we use the Moderate Growth scenario of Chapter 4. New orders would likely stimulate uranium industry expansion, particularly since supply arrangements for most new capacity would not be locked in by existing contracts. Figure 10-8 shows the results of this analysis. As in the preceding figure, foreign demand follows existing contractual supply arrangements until 1987, when actual reactor requirements begin to exceed those supplies; after that, the curve shows annual reactor requirements. In a similar fashion, all U.S. imports and uncovered needs are assumed met from imports in the near term, with half of long-term demand being supplied from abroad.

The result of this comparison suggests that uranium production capacity expansion could easily outstrip new demand growth, including major new international market participation by U.S. utilities. Moreover, much of the expansion would come from reserves where costs are significantly below those of existing mines. While it will be argued below that rigidities and sovereign interventions make major overall price declines in the primary market unlikely during much of the rest of this decade, the current overcapacity in the world uranium industry, and the potential for new low-cost production centers, makes cost considerations central to any long-term market assessment.

Assessing Costs

Estimates of production costs for existing and new production centers are invariably suspect, particularly when it is necessary to put such estimates for

Figure 10–8.   Demand on Non-U.S. Uranium Producers For Planned and Potential New Reactors (*Maximum Uranium Capacity Expansion*).

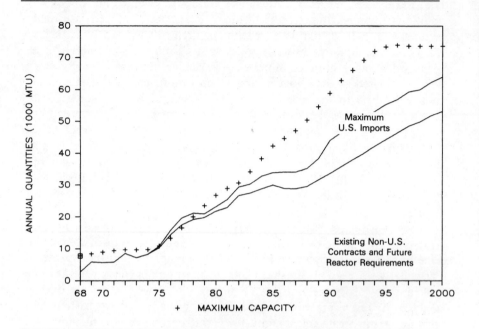

Note: Capacity expansion is that of Figure 10–6. Non-U.S. utilities are assumed to take deliveries under existing contracts until 1987, when new demand begins to emerge from under existing commitments; after this date, non-U.S. utilities are assumed to meet annual reactor needs through new purchases, but with no use of inventories existing at that time. U.S. utilities are assumed to continue deliveries under existing domestic and foreign contracts, to meet reported "uncovered needs" exclusively through imports, and to meet half of long-term needs through foreign purchases.

different nations on a comparable basis. A number of definitional issues associated with existing estimates were discussed in Chapter 3. These included the extent to which the "forward cost" concept used in official estimates can approximate either marginal cost or long-run resource cost and the difficulty of separating cost estimates from the way in which different reserve components in a given deposit are counted. In addition to these issues, there are problems that arise in assessing costs in particular situations.

In many past international mining ventures, capital investment was not made under open market conditions. Military procurement programs did not entail (or at least did not appear to entail) significant risks for developers; in some cases funds came directly from purchaser governments.

In the subsequent market slump, producer governments stepped in to maintain operations, protecting the basic industry investment. This pattern continued in the expanding international commercial market of the 1970s, when consumers driven by security of supply concerns effectively took on a significant share of development risk (thus subsidizing interest rates), either by direct investment or by signing long-term contracts that guaranteed returns to those providing investment funds. However, future uranium projects seem much more likely to involve relatively high risk premia, given weak market conditions and high project uncertainties. An exception to this may be those ventures involving producer government involvement, where much of the risk may be carried by government agencies. In Saskatchewan, for example, both provincial and federal agencies are taking major roles in new mining developments.

A second problem concerns accounting for infrastructural investment. Many new mines and mills are located in remote areas. These include mines in northern Saskatchewan, the Northern Territory of Australia, and Niger. The costs of developing infrastructure are generally attributed to the first mining ventures (unless they are picked up by government agencies), with subsequent new projects entailing smaller marginal infrastructural costs. For example, massive infrastructural investments were made in Niger, including a 600 km paved road to reach the Air Mountain uranium area. These costs were largely supported by the Somair and Cominak consortia, and the marginal infrastructural cost of new production would presumably be low. Unfortunately, major new prospects in Niger either have significantly lower ore grades and higher mining costs (Imouraren) or are likely to be delayed for logistic reasons (Arni being most naturally seen as a follow-on to Arlit).

A third problem arises in considering jointly produced minerals. Not only is there a question about assessment of "economic" reserves (where the amount of uranium estimated as economically recoverable may depend on assumptions about prices and other attributes of coproduced minerals), but also about how to allocate production costs. This situation arises to a minor degree for many uranium deposits but is important to exploitation in South Africa and central to future production at the massive Roxby Downs deposit in South Australia. For most South African mines, costs and profits attributed to uranium are generally small compared to those for gold. However, weak gold prices combined with rising mining costs may increase the importance of uranium as a coproduct. At Roxby Downs, uranium would account for up to half of the total market value of recovered minerals at recent depressed prices, though relative contributions to recovery costs are not yet clear.

In considering the market roles of various mining operations, at least in the short run, it is important to separate capital costs—most of which are now sunk costs—from operating costs and to distinguish between fixed and variable components of the latter. With surplus capacity and weak market conditions, operators of existing mines should be willing to produce as long as market prices are higher than operating costs (the capital investment having already been made and unrecoverable). The level of operation may depend on the relationship between fixed and variable costs and the nature of contracting and sales opportunities.

If fixed costs are high and prices are eroding, a producer will face a choice between shutting down or running near full capacity. If there is a base of long-term contracts that covers much of the fixed costs, and variable costs are low, the producer may choose to stockpile or make spot and other short-terms sales. This appears to be the situation at Rössing, where large capital investments are sunk and rather high fixed costs account for a significant fraction of operating costs. In contrast, Nabarlek (whose capital costs were small, but now equally irrelevant) appears to be much more flexible; with a higher proportion of variable costs, Nabarlek can easily alter output levels to meet sales requirements. For South African gold mining companies, fixed costs tend to be covered (at least according to current accounting methods) by gold operations, with uranium recovery largely a question of low variable costs. Mines in eastern Canada (those of Denison and Rio Algom) have high fixed and variable costs, while those in the west are generally characterized by fixed and variable costs that are relatively low, despite a physically inhospitable environment that raises labor and material costs.

Consideration of these factors suggests that it should be possible to give some rough ranking of comparative costs for different mining operations. Such assessments are quite subjective due to the difficulty of quantifying unique factors for each mine. The author's rankings are given in Table 10-1, where the "low" cost designation indicates basic costs (including capital recovery) below about US$20 (1983 dollars) per pound, "moderate" between $20 and $30, and "high" above $30 (but probably not greatly so). These cost estimates are idealized, as will be seen below.

Under idealized conditions, prices in a market with overcapacity should fall to the operating cost level for the mine whose output just satisfies the last unit of demand. Moreover, new mining ventures should be attractive only if *total* costs (including return on new investment) are below *operating* costs for the marginal existing mine. This would tend to limit new investment to mines with high ore grades, where both capital costs and operating costs are low. The capital and operating costs of uranium facilities are roughly proportional to the amount of ore-processing capacity, which will be low for mines with high ore grades. Such ventures would also be more flexible and adaptive in following rapid market changes.

These simple economic observations would be useful in predicting market trends if the international market functioned in this idealized way.

But while basic economic forces must eventually be felt, and while they help us in measuring departures from efficient behavior, there are important rigidities and market imperfections that must be examined if we are to understand the international uranium market. In subsequent sections we peel back the several overlays of contracting and other rigidities and government interventions that alter the nature and evolution of the market. We first examine the factors that shape the current market and that will affect it over the remainder of this decade.

MARKET RIGIDITIES AND SOVEREIGN CONSTRAINTS

As discussed above, if there were no long-term contracts or consumer equity commitments and no governments involved, consumers would be free to purchase from the supplier with the lowest price. Under competitive market conditions, that price would approach long-run cost (appropriately discounted and reflecting a small user cost). The discovery of new low cost reserves would result in changes in industry composition and existing market shares. In the near-term, those producers with existing mines and mills where operating costs were lower than total costs for new production would continue to sell, while those with higher operating costs would shut down. In the longer term, developers of new low-cost reserves would displace new capacity creation by those producers with higher costs.

As noted, these basic economic facts are ultimately likely to win out. However, the existing market not only has structural rigidities that pose high cost barriers to readjustment but also government involvements that inhibit actions that would otherwise alter market conditions. These rigidities and government interventions, in effect, change the resource cost structure and thus the effective supply curve. In general, most of these "imperfections" tend to hold the supply curve up; higher prices are necessary to bring forth a given increment of supply. In Australia, for example, discoveries of large low-cost reserves would have had a major effect in lowering the world supply curve, except for the government-imposed price floor, restraints on project development, and great uncertainties about future access to such reserves.

This effect is amplified by other uncertainties associated with government interventions. Investors in exploration or development of new reserves must take into account the possibility not only that the government of the host country will impose price or export restrictions (and most likely increase its share of the rents) but also that governments in other producer nations will alter pricing policies in such a way as to undermine market opportunities for the new producer. These possibilities increase the return required of new investment and may thus reinforce the existing price regime, rather than posing a threat to it.

While this situation might occur for other markets (as it does in the case of oil, for example), it is a particular problem for uranium in that demand is so inelastic and, at least for this decade, inventories so high. New low-cost producers that in other markets might find increased demand at lower prices will seldom find such opportunities in the uranium market; instead, such sellers will have to displace existing supply arrangements. This is difficult to do, for reasons discussed below.

In principle, demand for inventories might increase as prices drop. However, because deliveries under existing contracts have grown at the same time that expected reactor requirements have shrunk, inventory demand is largely saturated and in many cases is negative. That is, consumers have excess inventories that they are willing to sell or loan to other consumers in a market that is *perceived* as trending toward lower prices and greater security of supply. The net effect of inventory demand is thus to displace new primary demand, rather than adding to it.

In the following discussion, we consider the origins of barriers to readjustment—including contractual and equity obligations for both uranium and enrichment, and sovereign interventions. Effects on prices and market structure are of particular interest here. We then evaluate industry and market responses that might ease adjustment to altered market conditions.

Contracting and Equity Constraints

As discussed in Chapter 8, non-U.S. utility commitments to uranium and enrichment services have significantly exceeded reactor requirements over much of the history of nuclear power and, under existing contracts and equity commitments to mining and enrichment ventures, will continue to do so until late in this decade. The situation for uranium was shown in Figure 8–2 while the implications of enrichment contracts for uranium demand were explored in Figure 8–4.

Cumulative uranium inventories at this writing exceed four years forward supply and will continue to grow in absolute terms (though not in terms of years forward coverage because of nuclear capacity growth) until near the end of the decade unless efforts are made to alter scheduled deliveries (see Figure 8–3). The uranium procurement problem is aggravated by commitments to enrichment services, though these contractual imperatives begin to retreat several years before those associated with uranium procurement. There are thus dual obstacles to altering nuclear fuel procurements.

This situation gives rise to a number of forces. The first, obviously, is a growing incentive to reduce forward financial commitments to excess supplies of both uranium and enrichment services. This might be done through renegotiation of prices, or through termination of contracts. But here there are significant problems for at least some consumers.

Consumer supply commitments fall into three general categories. The first is a conventional contract in which a producer agrees to sell a given quantity of uranium or enrichment services to the consumer at a price set by an escalation formula reflecting a market price indicator, or by a producer government agency. On the expectation of sales, the producer invests in a mine and mill or an enrichment plant. In this case, the consumer may seek to end contracts or renegotiate prices down toward operating costs, with the owner writing off the sunk costs. The chief barriers inhibiting this are the contract itself, which may involve explicit or implicit penalties, or intervention by the producer government, as discussed below. In addition, utility incentives to renegotiate contracts may not be high, either because regulatory bodies allow costs under existing contracts to be passed through or because consumer governments put a premium on supply security, diversification, good relations with producers, or other nonprice objectives.

The second type of arrangement is similar to the first, except that contracts or other supply arrangements are the basis for project financing by third parties, one of which may be a consumer nation government agency or bank consortium. If these contracts are broken, or prices renegotiated downward, both the producer's profits and the financing agency's loans may be at risk. When the latter is an agency of the consumer utility's home government, there may be pressure for the utility to be modest in its demands. In effect, the utility (or its customers) may be asked to absorb part of the sunk costs, though there may still be room for contract price reductions if there are signicant profit margins.

This situation becomes more obvious in the third class of supply commitments, in which the buyer has invested directly in the production facility, though the accounting may become complicated. In such cases, the consumer's investment may be repaid through output at rate determined by the price set for the uranium or enrichment services. This price may include a capital recovery component plus an operating cost component, with the total price set by an escalation formula, or by a producer government agency. In principle, the consumer should make the same calculation as any producer, selling to itself as long as operating costs are lower than total costs of alternative supply. However, the details of the supply arrangement may affect both what happens as well as how it is reported.

For example, the utility may participate in the venture through a subsidiary (e.g., a member of a consortium in the producing nation) that bills the parent company at nominal price, including the capital component, while making profits or incurring losses itself. In this situation, the nature of utility regulation may play an important role: if the regulator allows the utility to pass along fuel costs to ratepayers—but not losses by a production subsidiary—there will be a strong incentive for the parent utility to continue to pay nominal contract prices for fuel deliveries and for its subsidiary to report higher revenues. Such an arrangement may result in higher total costs to the utility (for example, the subsidiary may have to pay higher taxes to producer nation governments), but also higher contributions to earnings. The net result may be that the mining venture will continue to recover sunk costs, shifting some of these costs to utility ratepayers.

These examples suggest that basic economic pressures toward lower prices in a market with significant surplus capacity may be held back by contractual rigidities of several types. There are also significant involvements by governments that alter market conditions and tend to inhibit readjustment.

Sovereign Intervention

Governments have always been significantly involved in international nuclear fuel markets. The history of this involvement was recounted in Chapter 2. Early motives for government participation included procurement of uranium and enrichment services for building nuclear weapons and sustenance of new industries perceived as important to national security or the economy. Later, concerns arose about nuclear weapons proliferation that led to greater government intervention in nuclear fuel trade. Some nations sought to use approval over conventional nuclear fuel trade as a source of leverage to influence the nuclear technology development and export activities of other nations; Canada and the United States were most conspicuous in this approach. Other national governments emphasized the value of assured supplies of conventional fuel in deterring nuclear developments that might threaten the nonproliferation regime. Some nations embraced both views. Following the first oil crisis, consumer nation governments took a wide range of actions to improve security of nuclear fuel supply for reactor programs perceived as essential to national security.

More recently, governments of uranium-producing nations have intervened (or threatened to do so) to promote the interests of domestic nuclear fuel industries in a weak international market. The United States has considered limits on imports of uranium, while simultaneously trying to hold onto its share of the international enrichment market. Other nations have attempted to sustain prices for existing contracts, put floors under or otherwise review new export prices, and retard market-threatening expansion. Still others have sought larger market shares for domestic producers or intensified the activities of state enterprises and agencies.

The ways in which various objectives and approaches enter into government actions—how they reinforce or conflict with each other—is different in each nation. In Part II we looked explicitly at the policy-formulation process in major uranium-producing nations. Here we summarize and develop the major themes for these and other nations participating in the international market and consider how particular forces affect market roles.

Australia.   Australian government policy toward uranium represents a close balancing of commercial and political factors—one that has not always been stable. The roles of these factors in determining policy, and their interactions, are more complex than they appear on the surface. Superficially, the political factors appear to arise largely from public concerns about nuclear power (currently focused largely on nuclear waste issues) and nuclear weapons proliferation, with additional concerns raised about the environmental impacts of uranium mining and its implications for aboriginal cultures. The obvious commercial issues have to do with attracting investment and earning the highest possible return on natural resource exports. These are, in fact, major issues.

But underlying these public issues are deeper domestic and foreign policy questions. There continue to be conflicts among the federal government, the states, and the Northern Territory over who owns, regulates, and benefits from natural resource development. There is even more critical dispute over which groups should benefit most from uranium exports. Aboriginal groups, labor organizations (of which there are many involved), private firms, and state and national governments all want a share of the rents. An immediate implication of this political fact is that prices must be high enough or volumes great enough to ensure sufficient rents to go around.

The rapidly shifting ground of Australian electoral politics reflects this imperative. For example, the Labor government elected in 1983 explicitly opposed uranium mining. Prior to that administration's election, labor groups and aboriginal organizations in the Northern Territory had frequently expressed opposition to uranium activities. After the election, the same or similar groups fought to sustain mining in the territory, undercutting support for the Labor party government. While there are undoubtedly noneconomic issues involved, economic factors are obviously extremely powerful.

These forces are greatest where ventures have already proceeded, or where there is promise of great wealth to be shared. For existing mines, such as Ranger, the distribution of substantial rents has created constituencies supporting the continuation of exports and efforts to maintain prices. If prices dropped, there would not only be less for some groups, but also political instability as groups tried to shift the burden of loss. The importance of this effect is enhaced by the preeminence of Ranger in the economy of the Northern Territory. In economically depressed South Australia, Roxby Downs promises thousands of new jobs and an influx of new business and wealth, with many of the same implications for state and national policy.

These domestic interests reinforce a general spirit of resource nationalism that asks full "value" for Australian uranium. Liberal-National Party government officials who instituted the requirement of a specific price floor for Australian exports argued that low spot market prices in the early 1980s did not fully reflect the long-term value of uranium. While economists might disagree with this proposition, many politicians would not. From the perspective of the latter, there are political costs involved in mining and exporting uranium. In addition to the scramble for rents and the need to please the resulting strong constituencies, political leaders must balance the interests of other groups. Here there are two problems.

The first arises from the need to maintain equity between regions. If uranium from South Australia can only be produced at higher cost than in the Northern Territory, it is difficult to retreat from regulations that would otherwise appear to give equal market opportunity to both regions. In addition, approval of new contracts at lower prices (where costs might allow) would undermine previous contract price terms, bringing demands for renegotiation and even the spectre of a net economic loss to the nation.

The second problem stems from the nuclear-related concerns of a significant fraction of the Australian public. If exports are to occur at all—which has at times been in question—they must be in exchange for substantial economic benefits to the nation. From this perspective, it is better to export a little at high prices, than to export a lot at lower prices. The balancing act here is difficult, particularly because the majority of the population resides in states without uranium resources.

The net effect of all of these factors appears to be a cautious approach to approving new mining ventures and a reluctance to alter uranium pricing regulations. Despite the fact that a more aggressive market strategy might yield long-term gains, it is unlikely that the current national government can pursue such a strategy. However, there is some chance that state and Northern Territory governments could act independently to increase export sales. The risk of this (apart from the question of national export approvals) is that it might aggravate public opposition to exports, destabilizing already fragile acceptance of some exports.

Canada.   Many of the domestic and international factors that influence Australian policy are also present in Canada. Contract approval and annual review of export prices are conducted by the national government, and there has been recurrent concern about nuclear weapons proliferation as well as serious interprovincial equity issues. The outcome in Canada, however, has been quite different than in Australia. Where Australian authorities have held firmly to an escalated floor price that is well above spot prices, Canadian officials have approved new contracts with prices tied to the spot market and even offered some discount from these prices. Where new ventures in Australia have had difficulty in attracting financing or obtaining government approval to proceed, new mines have gone ahead in Saskatchewan. And while domestic customers of Denison Mines in eastern Canada are effectively paying more than $50 per pound for some deliveries, foreign purchasers of western uranium are paying less than half this amount.

The different outcome in Canada appears to stem from how economic and political factors are handled. New mine development in Saskatchewan has a high level of participation by both provincial and national state enterprises—the Saskatchewan Mining Development Company and Eldorado. This has two effects. First, financing and risk are borne by governments, effectively lowering the threshold for proceeding and at times reducing costs. Second, participation by government agencies helps ensure both mining and export approvals. The balance of internal power is also somewhat different: at present, a larger set of forces has weakened federal authority and tipped the balance of power between the provinces and the federal government in the direction of the former.

Where historically the national government might have imposed export levies on western production to subsidize domestic consumption in the east—the case with oil—there does not appear to be adequate power to do so now for uranium. In addition, Ontario Hydro has provided the financing for expansion by Denison Mines and Rio Algom and is being repaid through output. As discussed in the preceding section, this arrangement introduces significant rigidities to contracting and pricing and helps preserve the status quo even under competitive threat from low-cost producers in the west.

The principal market for new low-cost producers in Saskatchewan is the United States. Sales have been made to both U.S utilities and uranium producers. The latter find such purchases a cheaper way to meet delivery commitments than through domestic production. Increases and elaborations of such arrangements will help reduce the pressure within the United States to impose restrictions on imports, though the possibility of such restrictions is a matter of concern to Canadian uranium producers. However, the fact that Canadian federal and provincial governments both have stakes in exports suggests that producers can expect government support in bilateral discussions and that stability of uranium trade may benefit from linkages to other important bilateral issues. With the United States as their major customer, Canadian producers also minimize the risk that proliferation concerns will escalate as a public issue threatening future exports.

African Producers.   Only in Niger and Gabon are governments directly involved in a significant way in uranium activities. Elsewhere, sovereign involvement in African uranium production is generally less apparent than that in either Canada or Australia. However, this does not mean that government is not relevant to the international market roles of South Africa and Namibia. Indeed, there the issue of sovereignty itself is fundamental to future uranium production and exports.

Uranium production and sales are affected by Niger's government in several ways. Through Onarem, the government participates directly in uranium production decisions and in export sales and has intervened to increase its own participation. The government also sets the official export price. However, this price is a nominal transfer price for the consortia operating in Niger. As discussed in the preceding section, such prices do not necessarily represent the true cost for the foreign firms involved in Niger. However, the official price, which is considerably higher than spot market prices, does affect tax revenues collected by Niger.

The role of Niger in international uranium markets is dominated by two facts. The first is the central importance of uranium revenues in Niger's economy. As a result, Niger must sell its share of output for as much as possible. Previously this has meant sales to Libya and other nations with little legitimate need for uranium; in the future, uranium from Niger may sell at higher than market prices to customers trying to avoid nonproliferation and other conditions imposed by some suppliers. For these reasons, and those noted above, it seems likely that the official "posted" price will be held consistently above spot market prices.

The second fact is that Niger's partners in uranium development are now largely agents of other governments, which in turn bring other non-market factors into uranium decisions. Private firms, the most prominent of which was the multinational Conoco, have mostly left Niger. The companies that remain include state enterprises or agencies from France, Spain, Italy, and Japan. The most important of these is the French governmental agency Cogema, which has increased the size of its position through takeovers (and bailouts) of other French firms.

Despite growing nationalism, France remains engaged in central Africa and seems likely to assist Niger in basic economic matters. In recent years, this has included purchases of uranium from Niger that were openly labeled noncommercial—forms of aid. However, such purchases feed into the French nuclear fuel system, which includes growing sales by Cogema of packages of uranium and enrichment services in other consumer nations, including the United States. Prices for such packages reflect not only the implicit aid subsidy to Niger, but also the enrichment surpluses resulting from France's lead role in Eurodif.

Cogema is also the principal agent in Gabon's uranium trade, an importance that has increased with the withdrawal of Union Carbide and other firms from exploration in the country. Nominal uranium prices appear to be lower than those of Niger, although given Gabon's close relationship with France, it is unclear what such prices really signify. Production has been reduced below capacity, matching reductions in exports which are now almost entirely to France. With petroleum its dominant source of income, Gabon is not as critically dependent on uranium as Niger and seems most likely to act in accord with France, rather than seeking a more aggressive independent market role.

In contrast to Niger and Gabon, the government of South Africa takes no direct role in uranium exports, except to require secrecy for such transactions. The industry is largely in the hands of a few major groups of gold mining firms. However, the government of South Africa does significantly influence that nation's role in the international market. Several consumer nations will not trade (at least publicly) with South Africa in protest of its apartheid policies. Others have favored purchasing uranium from South Africa because there are fewer restrictions on utilization and retransfers than for purchases from Australia, Canada, or the United States. However, it is difficult to demonstrate that South Africa is able to charge higher prices as a result of this policy—there are too many other factors affecting prices to make such a determination.

The major government impact on uranium mining and sales is indirect. Because of the need to retain skilled workers and maintain efficient output in gold and uranium production, the mining houses have sometimes found themselves allied with liberal elements in the apartheid struggle. Government efforts to limit black rights, including the homelands policy, tend to reduce productivity and limit economic contributions of black workers. In addition, and perhaps more seriously, the continuing confrontation over racial issues aggravates conflicts in the mines over relative wage scales and job assignments for white and black workers.

The perceived potential instability of the South African regime also affects the type of market roles possible to South African producers. Short-term contracts and spot sales are more common than long-term agreements, exposing producers to rapid changes in prices and revenues and altering the investment and production environment. Coupled with the relatively low cost of uranium as a coproduct with gold (low in part because of accounting and government tax rules), this can result in significant inventory building in soft market periods.

Inventories were a source of large profits to South African mining firms in the panic of the mid-1970s. However, they may play a different role in the future, overhanging the market and rapidly adding to spot market availability of material when demand increases and prices begin to rise. This problem will increase as more production capacity emerges from under existing contractual export commitments. How it will be handled is not yet clear, though it seems likely that some action will be taken to restrain excessive production and alleviate competitive pressures.

Such actions have been taken for coal and other exports, with allocation of market shares to individual producers. This was also the case for uranium prior to the great upsurge in prices in the 1970s when Nufcor—formerly a government agency but now run collectively by the mining houses—acted as sole sales agent and allocated new contracts among mines. A return to this approach for uranium exports would not require government intervention, though one might well argue that it would be an indirect result of sovereign uncertainty since long-term contracts would otherwise not leave such a large fraction of the industry exposed. The effect would be the same as if the government intervened: an increase in market concentration and the potential for collusion.

The problem of political uncertainty arises in even more exaggerated form in Namibia. Because of low ore grades and high fixed costs, the Rössing mine requires a baseload of long-term contracts to continue operation. Existing contracts would provide this baseload. However, consumers with excess supply commitments are under pressure to reduce these commitments, or at least to renegotiate prices. Because of sanctions against trade with Namibia, uncertainty about future political stability, and little likelihood of major price reductions, some customers may regard breaking contracts with Rössing as less costly—in political as well as economic terms—than reducing other supply commitments. The result may be either reduced output in Namibia, or stockpiling for future sales. If long-term contract reductions are not too great, it may be possible to cover fixed costs and continue full output with stocks being used for future spot market sales.

In the longer term, it seems likely that Rössing will continue to play an important role in the world market, as it is a valuable asset to the Namibian economy and to whatever government eventually prevails. The operators of Rössing have made efforts to distance themselves from direct connections with South Africa and to maintain communications with dissident groups. However, this is a difficult situation and one where international political costs and uncertainties about future governance cannot help but restrict Namibia's role in international markets, until such stresses are resolved.

The United States.   The U.S. government is still wrestling with the implications of its changing relationship to international nuclear fuel markets. As discussed in the preceding chapter, potential low-cost supplies—especially from Canada—threaten to render many U.S. uranium operations noncompetitive; because U.S. utilities are less constrained by long-term contractual rigidities than non-U.S. utilities, the United States is the potential target of marketing efforts by owners of surplus production capacity around the globe as well as by new low-cost mines in Saskatchewan.

The effect of U.S. imports on the world market would be to increase demand and support higher prices than would otherwise obtain. The equilibrium price for an integrated world market would be between that for the international market without the United States and the higher market-clearing price for the United States taken in isolation. Thus, one should not take past world spot prices as necessarily the measure of prices at which larger quantities of imports would be available to U.S. utilities. Indeed, non-U.S. suppliers may anticipate this effect and price uranium accordingly, either through specific price provisions or by coupling prices to spot market quotes or other indicators that will automatically rise as U.S. entry increases.

The possibility of U.S. import restrictions will provide additional incentives to non-U.S. producer governments to resist price reductions that would threaten greater imports by the United States and trigger protectionist actions that would eventually lead to lower prices for worldwide sales. This argument depends on the existence of a common perception on the part of major suppliers. Otherwise, it may be in the interest of one supplier to undersell others in an effort to capture a larger share of the U.S market. A producer nation with strong bilateral relationships to the United States—and low uranium production costs—will be in the strongest position to cut prices to increase market share, since the U.S. government will be less likely to oppose such imports from a close ally and trading partner. Canada is the most obvious candidate for this role.

While there is concern on the part of some domestic uranium producers, and their representatives in Congress, the U.S. government is in a somewhat different situation regarding enrichment. As discussed in Chapter 1, the United States, in only a decade, has moved from being a monopoly supplier to holding a rapidly shrinking, minority market share. The difficult situation faced by the United States affects both enrichment and uranium markets. Most U.S. enrichment exports come from underutilized plants originally built for weapons purposes. These investments, and the capital costs of subsequent upgradings, are sunk.

Like the owners of existing uranium mines, enrichment plant owners should continue to sell as long as prices are higher than operating costs. Competitive prices, based on operating costs, would be lower than current, and expected, official U.S. prices. However, this economic reality is difficult to accept for a number of political as well as budgetary reasons. Like miners with long-term contracts, the near-term impulse is to hold to the rigidities of existing contractual commitments—and to high prices, even though this may mean the longer term erosion of market share.

The enrichment imperative also affects the uranium market. With low uranium prices and high enrichment prices, utilities would like to increase the tails assay in the enrichment process, reducing the enrichment services purchased and increasing uranium delivered to the enrichment plant. If U.S. enrichment contracting makes it costly to increase the tails assay, as seems likely, uranium demand will be lower than otherwise. However, this effect will probably be significantly outweighed by continued difficulty on the part of consumers in reducing enrichment commitments. Utilities with enrichment contracts greater than their needs would like to reduce these commitments. If U.S. enrichment policy made it easy for utilities to cancel or otherwise reduce enrichment commitments, demand on the uranium market—which is driven in part by enrichment contracting—would also be reduced. This also seems unlikely and U.S. government efforts to preserve existing enrichment commitments will, on balance, continue to sustain uranium demand.

STRUCTURAL ADJUSTMENT AND MARKET EVOLUTION

The preceding two sections have portrayed the international uranium market as a system in which existing and potential supply are significantly in excess of demand but where there are also important rigidities arising from contractual and governmental constraints. The latter inhibit market adjustment and give rise to economic inefficiencies. Prices and production costs are higher than they would be without constraints, excessive inventories have accumulated, and new long-term investment in low-cost production is being discouraged. The last of these helps perpetuate inefficiencies.

Despite this, adjustments have begun to occur, though not all of the changes now visible can be regarded as desirable. Some contracts are being renegotiated, reversing the price increases made in the crisis years of the last decade and stretching out or otherwise altering delivery schedules. Such revisions are most noticeable in the United States, for reasons discussed above and in Chapter 9.

Compensating for rigidities upstream, secondary markets have grown rapidly. Consumers forced to take unwanted enrichment services or uranium are reselling or loaning the excess, reducing the size of their loss. In the United States, secondary market volumes have grown to rival primary procurements. Indeed, some mining companies find it cheaper to purchase or even repurchase uranium sold to utilities to meet current delivery commitments rather than mining new ore.

Outside the United States, secondary markets have been slower to develop, in part because few consumers had near-term requirements and in part because of the nature of the supply arrangements. The most rapid growth in secondary sales is for enrichment services, sometimes coupled with uranium in package deals, with a significant fraction of these sales to U.S. utilities. The earliest of these rearrangements resulted from reactor deferrals or cancellations, an example being uranium from Rössing originally enriched in the Soviet Union for Austria's Tullnerfeld reactor being sold to Rochester Gas and Electric in 1980.

The increasingly rich elaboration of secondary markets—with traditional utility and producer participants acting in new ways and assisted by a variety of brokers and financial participants—is an example of structural adaptation and compensation for failure of the primary market to adjust to changed circumstances. It does not, however, greatly reduce the efficiency losses already incurred, but only redistributes the gains and losses.

Other changes resulting from overexpansion are more worrisome. There has been significant retrenchment in the uranium industry outside the United States, with the victims largely private firms with new projects to develop. Conoco has retreated from Imouraren in Niger and from activities in the United States, Esso from Yeelirrie and the parent Exxon from operations in the United States and Canada, Gulf from Rabbit Lake and other prospects in Saskatchewan as well as from Mt. Tayor in the United States, and Getty is stalled at Jabiluka, despite its richness. Rio Tinto Zinc has closed down high cost operations in Australia, and must be reconsidering its role in Namibia. Of the large "private" companies, only Uranertz (at Key Lake) and British Petroleum and Western Mining (at Roxby Downs) are still in the development game, and of these BP is partially owned by the government of Great Britain and Uranerz is closely connected to the West German utility RWE.

Instead of private companies, state enterprises and consortia representing consumer groups now dominate the international industry. In Canada, SMDC and Eldorado are now in charge (except for the high-cost operations of Denison and Rio Algom in the east); Cogema has more fully consolidated within itself French procurement activities, expanded its activities (sometimes by default, as in Africa, but elsewhere, as in the United States, Canada and Australia, by design), and undertaken to expand sales of both uranium and enrichment services worldwide; and in Niger, Onarem will have to sell most of that nation's uncommitted production. Even in the new producing nations—Argentina, Brazil, Mexico, Portugal, and Spain—state enterprises dominate.

These changes may ensure that investment goes ahead where it might otherwise not, due to existing rigidities that undermine the promise of new investment in low-cost supplies. However, they also mean an increase in market concentration, reversing some of the gains in market evolution of the last decade and reducing the forces of competition. Even where state enterprises have not taken over, effective market concentration has increased: in South Africa, Nufcor will likely again act as sole sales agent and, in Australia, government is restraining industry development by refusing permission for several promising ventures and effectively eliminating price competition through the contract and price approval process.

These structural changes indicate the difficulty of modeling the international uranium market, and of making predictions of prices, trade patterns, and other aspects of future markets. Conventional competitive equilibrium models have failed to explain adequately many international commodity markets, including grain and coal—market imperfections appear to play essential roles. Efforts to include such imperfections have focused on characterizing and exploring the implications of various actors exercising market power. This has proven difficult to do for markets involving fewer rigidities and sovereign involvements than that for uranium. And, as our earlier discussion suggests, not all uranium market actors can convincingly be expected to pursue rational economic maximization. For example, the level of Australian exports seems to reflect political concerns about public approval of exports above some particular level. Moreover, such factors may change rapidly as, for example, the likely reduction in approval following the next nuclear weapons proliferation event.

Such internally oriented factors may have effects similar to those of externally oriented wealth-maximizing strategies, and thus be difficult to distinguish. If such factors were static, it would not be necessary to make distinctions, as all objectives of producers or consumers might be represented effectively by implicit tariffs for exports or imports. Unfortunately, markets are not static and predictive power will be enhanced only by understanding the origin of such tariffs and the forces for change.

These observations bear on the perennial question of potential cartelization of the international uranium market. The current picture of market concentration and the presence of governments obviously raises this possibility, just as it did in the early 1970s. But, just as then, it may be difficult to demonstrate effective cartelization, even if there is evidence of collusion (which there was in 1972–74, but which seems much less likely today). As our review suggests, there are many reasons for prices to be different from long-run resource cost, because of special conditions in either producer or consumer nations, and often because of both.

Over the remainder of this decade, contractual rigidities seem likely to endure, the artifact of earlier consumer desires for diversification and supply security, as well as of transient market power for producers. While some of these agreements may be broken or prices pushed downward, prices are not likely to fall to the level of long-run cost. A more accurate measure of the latter may come from prices in the secondary and spot markets. Purchasers in these markets will be predominantly U.S. buyers for the next few years, perhaps paying less on average than many foreign utilities—a massive irony, given the high cost of producing from U.S. domestic reserves.

In the longer term the world market will have to accommodate to the geologic fact of very different resource pools: the historic high cost reserves of sandstones and quartz-pebble conglomerates, the low cost unconformity-related reserves of Canada and Australia, and new kinds of deposits like the large moderate-grade deposit at Roxby Downs. Current indications are that there may be many more deposits like Key Lake and Jabiluka, though at greater depths and with no evident outcroppings. Greater depth and higher exploration costs than for existing mines will mean marginally higher uranium costs. And as these costs rise, there will be competition from deposits like Roxby Downs, where coproduction and economies of scale will add large new expanses to the cumulative uranium resource supply curve without great increases in costs.

This optimism about the natural resources on which nuclear power depends must be tempered by concern about how smooth the evolutionary path will be, and about the basic future of nuclear power itself. It is unlikely that there will be no new disruptive events in the international nuclear fuel market: there are only two uses for uranium and nuclear technology, and the tension between peaceful and nonpeaceful uses will remain an essential factor in international relations and in the domestic politics of key supplier nations. In fact, current surpluses, commercial pressures—and what seems to be a lower ranking of nonproliferation on the foreign policy agendas of many nations—may be increasing the likelihood of disruptive events.

For much of the remainder of this decade, there are few events that could radically change international nuclear fuel markets. Excess production capacity, huge stocks, and active secondary markets could compensate even for the loss of a major supplier from the world market. While there might be transient increases in prices, there would not be the kind of security and price panic that afflicted the immature market of the mid-1970s. However, as market adjustment occurs—as high cost production capacity is eliminated and stocks are drawn down—vulnerability to events of non-market origin may again increase, as may price volatility generally.

The nature of the future market, and its sensitivity to disruption, will depend on several factors. First, it is fairly obvious that increased demand on the international market from the United States will hasten the process of adjustment of supply and demand. This could increase or decrease market vulnerability. If new demand and higher prices stimulated widespread new investment in exploration and development of low-cost supplies whose output might be rapidly scaled up, the effect would be to make the market not only more efficient but also more resilient. However, if U.S. entry simply absorbs excess supply from a rigid market system, it will only increase the importance of those rigidities in market clearing under abnormal conditions.

The second major factor is the nature of investment. As we have seen, most past international investment has been tied directly to trade agreements between consumer and producer. This followed in the tradition established by the weapons states in their procurements, with energy security or postcolonial relationships replacing military security as the driving force. It seems unlikely that this can continue. Consumer nations are now suffering the high prices, oversupply, and other consequences of accepting large shares of risks under such past agreements.

Moreover, most near-term demand will come from U.S. utilities that are quite unlikely to invest in joint ventures abroad. Instead, these utilities are likely either to pursue a spot purchase strategy or to enter only into longer-term contracts that are tied to spot prices, in preference to traditional contracts specifying prices in advance. This consumer preference would reinforce the market picture drawn above—with most sales coming from existing mines or secondary markets, with little new investment. Where new investment does occur, it is more likely to come from state enterprises rather than private firms. With a significant component of risk coming from the state itself, state enterprises may see a different risk structure than would private firms. And in many cases, such organizations accept risks more easily than private companies.

This investment pattern has important implications for uranium market conditions in the 1990s and beyond. High perceived risks that undermine investment will result in higher long-run prices. Investment by state enterprises, as in Saskatchewan, will reduce this effect—bringing more low-cost reserves to market, but at the expense of a higher level of sovereign involvement in the international market and perhaps greater instability in the next decades. This tradeoff can hardly be attractive to consumers concerned about long-run supply security.

The most important factor affecting the international market is the outlook for nuclear power itself. Current market conditions result in part from the worldwide collapse of nuclear power growth expectations. As we have seen, existing uranium production capacity exceeds reactor requirements for all plants in operation, under construction or on order. While there may be minor shifts from high to low-cost reserves, major new investment in uranium will depend significantly on the level of new orders for nuclear power plants. If such orders do not materialize, the world uranium industry can look forward to a prolonged depression, punctuated, perhaps, by transient disruptions arising from national and international policy shifts.

If significant numbers of new plants are ordered—an unlikely prospect—new uranium investment should follow. While there has been concern in the uranium industry that future new demand could not be met promptly, this seems unlikely. Known reserves are sufficient for massive expansion of production capacity and such expansion would proceed quickly under realistic economic stimulus. Twice before in its history, the industry has been called upon to increase output by great amounts over short periods of time—first under the stimulus of weapons procurements and then during the demand panic of the 1970s. In both cases, physical capacity and output expanded smoothly and rapidly, despite the immaturity of the industry and major policy obstacles. Today, the industry is better prepared than ever to meet growing demand, and even to cope with rapid changes in the policy environment.

APPENDIX A

SUPPLIERS' GUIDELINES

**Note for Transmission of Nuclear Suppliers
Guidelines to IAEA**

The Permanent Mission of the United States of America presents its compliments to the Director General of the International Atomic Energy Agency and has the honor to enclose copies of three documents which have been the subject of discussion between the Government of the United States of America and a number of other governments.

The Government of the United States of America has decided that, when considering the export of nuclear material, equipment or technology, it will act in accordance with the principles contained in the attached documents.

In reaching this decision, the Government of the United States of America is fully aware of the need to contribute to the development of nuclear power in order to meet world energy requirements, while avoiding contributing in any way to the dangers of a proliferation of nuclear weapons or other nuclear explosive devices, and of the need to remove safeguards and nonproliferation assurances from the field of commercial competition.

The Government of the United States of America hopes that other governments may also decide to base their own nuclear export policies upon these documents.

The Government of the United States of America requests that the Director-General of the International Atomic Energy Agency circulate the texts of this note and its enclosures to all member governments for their information and as a demonstration of support by the Government of the United States of America for the Agency's nonproliferation objectives and safeguards activities.

Nuclear Suppliers Group Guidelines for Nuclear Transfers

1. The following fundamental principles for safeguards and export controls should apply to nuclear transfers to any non–nuclear-weapon state for peaceful purposes. In this connection, suppliers have defined an export trigger list and agreed on common criteria for technology transfers.

Prohibition on Nuclear Explosives

2. Suppliers should authorize transfer of items identified in the trigger list only upon formal governmental assurances from recipients explicitly excluding uses which would result in any nuclear explosive device.

Physical Protection

3. (a) All nuclear materials and facilities identified by the agreed trigger list should be placed under effective physical protection to prevent unauthorized use and handling. The levels of physical protection to be ensured in relation to the type of materials, equipment and facilities, have been agreed by suppliers, taking account of international recommendations.

   (b) The implementation of measures of physical protection in the recipient country is the responsibility of the government of that country. However, in order to implement the terms agreed upon amongst suppliers, the levels of physical protection on which these measures have to be based should be the subject of an agreement between supplier and recipient.

   (c) In each case special arrangements should be made for a clear definition of responsibilities for the transport of trigger list items.

Safeguards

4. Suppliers should transfer trigger list items only when covered by IAEA safeguards, with duration and coverage provisions in conformance with the GOV/1621 guidelines. Exceptions should be made only after consultation with the parties to this understanding.

5. Suppliers will jointly reconsider their common safeguards requirements, whenever appropriate.

Safeguards Triggered by the Transfer of Certain Technology

6. (a) The requirements of paragraphs 2, 3, and 4 above should also apply to facilities for reprocessing, enrichment, or heavy water production, utilizing technology directly transferred by the supplier or derived from transferred facilities, or major critical components thereof.

   (b) The transfer of such facilities, or major critical components thereof, or related technology, should require an undertaking (1) that IAEA safeguards apply to any facilities of the same type (i.e., if the design, construction or operating processes are based on the same or similar physical or chemical processes, as defined in the trigger list) constructed during an agreed period in the recipient country and (2) that there should at all times be in effect a safeguards agreement permitting the IAEA to apply Agency safeguards with respect to such facilities identified by the recipient, as using transferred technology.

Special Controls on Sensitive Exports

7. Suppliers should exercise restraint in the transfer of sensitive facilities, technology and weapons-usable materials. If enrichment or reprocessing facilities, equipment or technology are to be transferred, suppliers should encourage recipients to accept, as an alternative to national plants, supplier involvement and/or other appropriate multinational participation in resulting facilities. Suppliers should also promote international (including IAEA) activities concerned with multinational regional fuel cycle centers.

Special Controls on Export of Enrichment Facilities, Equipment and Technology

8. For a transfer of an enrichment facility, or technology therefore, the recipient nation should agree that neither the transferred facility, nor any facility based on such technology, will be designed or operated for the production of greater than 20% enriched uranium without the consent of the supplier nation, of which the IAEA should be advised.

Controls on Supplied or Derived Weapons-Usable Material

9. Suppliers recognize the importance, in order to advance the objectives of these guidelines and to provide opportunities further to reduce the risks of proliferation, of including in agreements on supply of nuclear materials

or of facilities which produce weapons-usable material, provisions calling for mutual agreement between the supplier and the recipient on arrangements for re-processing, storage, alteration, use, transfer or retransfer of any weapons-usable material involved. Suppliers should endeavor to include such provisions whenever appropriate and practicable.

Controls on Retransfer

10. (a) Suppliers should transfer trigger list items, including technology defined under paragraph 6, only upon the recipient's assurance that in the case of:

(1) retransfer of such items, or

(2) transfer of trigger list items derived from facilities originally transferred by the supplier, or with the help of equipment or technology originally transferred by the supplier;

the recipient of the retransfer or transfer will have provided the same assurances as those required by the supplier for the original transfer.

(b) In addition the supplier's consent should be required for; (1) any retransfer of the facilities or major critical components, or technology described in paragraph 6; (2) any transfer of facilities or major critical components derived from those items; (3) any retransfer of heavy water or weapons-usable material.

Supporting Activities

Physical Security

11. Suppliers should promote international co-operation on the exchange of physical security information, protection of nuclear materials in transit, and recovery of stolen nuclear materials and equipment.

Support for Effective IAEA Safeguards

12. Suppliers should make special efforts in support of effective implementation of IAEA safeguards. Suppliers should also support the Agency's efforts to assist member states in the improvement of their national systems of accounting and control of nuclear material and to increase the technical effectiveness of safeguards.

Similarly, they should make every effort to support the IAEA in increasing further the adequacy of safeguards in the light of technical developments and the rapidly growing number of nuclear facilities, and to support appropriate initiatives aimed at improving the effectiveness of IAEA safeguards.

TREATY ON THE NON-PROLIFERATION
OF NUCLEAR WEAPONS

The States concluding this Treaty, hereinafter referred to as the "Parties to the Treaty,"

Considering the devastation that would be visited upon all mankind by a nuclear war and the consequent need to make every effort to avert the danger of such a war and to take measures to safeguard the security of peoples,

Believing that the proliferation of nuclear weapons would seriously enhance the danger of nuclear war,

In conformity with resolutions of the United Nations General Assembly calling for the conclusion of an agreement on the prevention of wider dissemination of nuclear weapons,

Undertaking to cooperate in facilitating the application of International Atomic Energy Agency safeguards on peaceful nuclear activities,

Expressing their support for research, development and other efforts to further the application, within the framework of the International Atomic Energy Agency safeguards system, of the principle of safeguarding effectively the flow of source and special fissionable materials by use of instruments and other techniques at certain strategic points,

Affirming the principle that the benefits of peaceful applications of nuclear technology, including any technological by-products which may be derived by nuclear-weapons States from the development of nuclear explosive devices, should be available for peaceful purposes to all Parties to the Treaty, whether nuclear-weapon or non-nuclear-weapon States,

Convinced that, in furtherance of this principle, all Parties to the Treaty are entitled to participate in the fullest possible exchange of scientific information for, and to contribute alone or in cooperation with other States to, the further development of the applications of atomic energy for peaceful purposes,

Declaring their intention to achieve at the earliest possible date the cessation of the nuclear arms race and to undertake effective measures in the direction of nuclear disarmament,

Urging the cooperation of all States in the attainment of this objective,

Recalling the determination expressed by the Parties to the 1963 Treaty banning nuclear weapon tests in the atmosphere in outer space and under water[1] in its Preamble to seek to achieve the discontinuance of all test explosions of nuclear weapons for all time and to continue negotiations to this end,

---

1. TIAS 5433; 14 UST 1313.

Sensitive Plant Design Features

13. Suppliers should encourage the designers and makers of sensitiv
ment to construct it in such a way as to facilitate the applica
safeguards.

Consultations

14. (a) Suppliers should maintain contact and consult through
channels on matters connected with the implementation
guidelines.

    (b) Suppliers should consult, as each deems appropriate, wi
Governments concerned on specific sensitive cases, to ens
any transfer does not contribute to risks of conflict or ins

    (c) In the event that one or more suppliers believe that there ha
violation of supplier/recipient understandings resulting fr
guidelines, particularly in the case of an explosion of a
device, or illegal termination or violation of IAEA safegua
recipient, suppliers should consult promptly through di
channels in order to determine and assess the reality and
the alleged violation.

Pending the early outcome of such consultations, suppliers
act in a manner that could prejudice any measure that
adopted by other suppliers concerning their current cont
that recipient.

Upon the findings of such consultations, the suppliers, b
mind Article XII of the IAEA Statute, should agree on a
priate response and possible action which could include
mination of nuclear transfer to that recipient.

15. In considering transfers, each supplier should exercise prudenc
regard to all the circumstances of each case, including any risk t
nology transfers not covered by paragraph 6, or subsequent ret
might result in unsafeguarded nuclear materials.

16. Unanimous consent is required for any changes in these guide
cluding any which might result from the reconsideration ment
paragraph 5.

21 September 1977

Desiring to further the easing of international tension and the strengthening of trust between States in order to facilitate the cessation of the manufacture of nuclear weapons, the liquidation of all their existing stockpiles, and the elimination from national arsenals of nuclear weapons and the means of their delivery pursuant to a treaty on general and complete disarmament under strict and effective international control,

Recalling that, in accordance with the Charter of the United Nations,[1] States must refrain in their international relations from the threat or use of force against the territorial integrity or political independence of any State, or in any other manner inconsistent with the Purposes of the United Nations, and that the establishment and maintenance of international peace and security are to be promoted with the least diversion for armaments of the world's human and economic resources,

Have agreed as follows:

Article I

Each nuclear-weapon State Party to the Treaty undertakes not to transfer to any recipient whatsoever nuclear weapons or other nuclear explosive devices or control over such weapons or explosive devices directly, or indirectly; and not in any way to assist, encourage, or induce any non–nuclear-weapon state to manufacture or otherwise acquire nuclear weapons or other nuclear explosive devices, or control over such weapons or explosive devices.

Article II

Each non–nuclear-weapon State Party to the Treaty undertakes not to receive the transfer from any recipient whatsoever of nuclear weapons or other nuclear explosive devices or control over such weapons or explosive devices directly, or indirectly; not to manufacture or otherwise acquire nuclear weapons or other nuclear explosive devices; and not to seek or receive any assistance in the manufacture of nuclear weapons or other nuclear explosive devices.

Article III

1. Each non–nuclear-weapon State Party to the Treaty undertakes to accept safeguards, as set forth in an agreement to be negotiated and concluded

1. TS 993; 59 Stat. 1031.

with the International Atomic Energy Agency in accordance with the Statute of the International Atomic Energy Agency[1] and the Agency's safeguards system, for the exclusive purpose of verification of the fulfillment of its obligations assumed under this Treaty with a view to preventing diversion of nuclear energy from peaceful uses to nuclear weapons or other nuclear explosive devices. Procedures for the safeguards required by this article shall be followed with respect to source or special fissionable material whether it is being produced, processed or used in any principal nuclear facility or is outside any such facility. The safeguards required by this article shall be applied on all source or special fissionable material in all peaceful nuclear activities within the territory of such State, under its jurisdiction, or carried out under its control anywhere.

2. Each State Party to the Treaty undertakes not to provide: (a) source or special fissionable material, or (b) equipment or material especially designed or prepared for the processing, use or production or special fissionable material, to any non–nuclear-weapon state for peaceful purposes, unless the source or special fissionable material shall be subject to the safeguards required by this article.

3. The safeguards required by this article shall be implemented in a manner designed to comply with article IV of this Treaty, and to avoid hampering the economic or technological development of the Parties or international cooperation in the field of peaceful nuclear activities, including the international exchange of nuclear material and equipment for the processing, use or production of nuclear material for peaceful purposes in accordance with the provisions of this article and the principle of safeguarding set forth in the Preamble of the Treaty.

4. Non–nuclear-weapon States Party to the Treaty shall conclude agreements with the International Atomic Energy Agency to meet the requirements of this article either individually or together with other States in accordance with the Statute of the International Atomic Energy Agency. Negotiation of such agreements shall commence within 180 days from the original entry into force of this Treaty. For States depositing their instruments of ratification or accession after the 180-day period, negotiation of such agreements shall commence not later than the date of such deposit. Such agreements shall enter into force not later than eighteen months after the date of initiation of negotiations.

Article IV

1. Nothing in this Treaty shall be interpreted as affecting the inalienable right of all the Parties to the Treaty to develop research, production and

1. TIAS 3873; UST 1003.

use of nuclear energy for peaceful purposes without discrimination and in conformity with articles I and II of this Treaty.

2. All the Parties to the Treaty undertake to facilitate, and have the right to participate in, the fullest possible exchange of equipment, materials and scientific and technological information for the peaceful uses of nuclear energy. Parties to the Treaty in a position to do so shall also cooperate in contributing alone or together with other States or international organizations to the further development of the applications of nuclear energy for peaceful purposes, especially in the territories of non–nuclear-weapon States Party to the Treaty, with due consideration for the needs of the developing areas of the world.

Article V

Each Party to the Treaty undertakes to take appropriate measures to ensure that, in accordance with this Treaty, under appropriate international observation and through appropriate international procedures, potential benefits from any peaceful applications of nuclear explosions will be made available to non–nuclear-weapon States Party to the Treaty on a non-discriminatory basis and that the charge to such Parties for the explosive devices used will be as low as possible and exclude any charge for research and development. Non–nuclear-weapon States Party to the Treaty shall be able to obtain such benefits, pursuant to a special international agreement or agreements, through an appropriate international body with adequate representation of non–nuclear-weapon States. Negotiations on this subject shall commence as soon as possible after the Treaty enters into force. Non–nuclear-weapon States Party to the Treaty so desiring may also obtain such benefits pursuant to bilateral agreements.

Article VI

Each of the Parties to the Treaty undertakes to pursue negotiations in good faith on effective measures relating to cessation of the nuclear arms race at an early date and to nuclear disarmament, and on a treaty on general and complete disarmament under strict and effective international control.

Article VII

Nothing in this Treaty affects the right of any group of States to conclude regional treaties in order to assure the total absence of nuclear weapons in their respective territories.

Article VIII

1. Any Party to the Treaty may propose amendments to this Treaty. The text of any proposed amendment shall be submitted to the Depositary Governments which shall circulate it to all Parties to the Treaty. Thereupon, if requested to do so by one-third or more of the Parties to the Treaty, the Depositary Governments shall convene a conference, to which they shall invite all the Parties to the Treaty, to consider such an amendment.

2. Any amendment to this Treaty must be approved by a majority of the votes of all the Parties to the Treaty, including the votes of all nuclear-weapon States Party to the Treaty and all other Parties which, on the date the amendment is circulated, are members of the Board of Governors of the International Atomic Energy Agency. The amendment shall enter into force for each Party that deposits its instrument of ratification of the amendment upon the deposit of such instruments of ratification by a majority of all the Parties, including the instruments of ratification of all nuclear-weapon States Party to the Treaty and all other Parties which, on the date the amendment is circulated, are members of the Board of Governors of the International Atomic Energy Agency. Thereafter, it shall enter into force for any other Party upon the deposit of its instrument of ratification of the amendment.

3. Five years after the entry into force of this Treaty, a conference of Parties to the Treaty shall be held in Geneva, Switzerland, in order to review the operation of this Treaty with a view to assuring that the purposes of the Preamble and the provisions of the Treaty are being realized. At intervals of five years thereafter, a majority of the Parties to the Treaty may obtain, by submitting a proposal to this effect to the Depositary Governments, the convening of further conferences with the same objective of reviewing the operation of the Treaty.

Article IX

1. This Treaty shall be open to all States for signature. Any State which does not sign the Treaty before its entry into force in accordance with paragraph 3 of this article may accede to it at any time.

2. This Treaty shall be subject to ratification by signatory States. Instruments of ratification and instruments of accession shall be deposited with the Governments of the United States of America, the United Kingdom of Great Britain and Northern Ireland and the Union of Soviet Socialist Republics, which are hereby designated the Depositary Governments.

3. This Treaty shall enter into force after its ratification by the States, the Governments of which are designated Depositaries of the Treaty, and forty other States signatory to this Treaty and the deposit of their instruments of ratification. For the purposes of this Treaty, a nuclear-weapon State is one which has manufactured and exploded a nuclear weapon or other nuclear explosive device prior to January 1, 1967.

4. For States whose instruments of ratification or accession are deposited subsequent to the entry into force of this Treaty, it shall enter into force on the date of the deposit of their instruments of ratification or accession.

5. The Depositary Governments shall promptly inform all signatory and acceeding States of the date of each signature, the date of deposit of each instrument of ratification or of accession, the date of the entry into force of this Treaty, and the date of receipt of any requests for convening a conference or other notices.

6. This Treaty shall be registered by the Depositary Governments pursuant to article 102 of the Charter of the United Nations.

Article X

1. Each Party shall in exercising its national sovereignty have the right to withdraw from the Treaty if it decides that extraordinary events, related to the subject matter of this Treaty, have jeopardized the supreme interests of its country. It shall give notice of such withdrawal to all other Parties to the Treaty and to the United Nations Security Council three months in advance. Such notice shall include a statement of the extraordinary events it regards as having jeopardized its supreme interests.

2. Twenty-five years after the entry into force of the Treaty, a conference shall be convened to decide whether the Treaty shall continue in force indefinitely, or shall be extended for an additional fixed period or periods. This decision shall be taken by a majority of the Parties to the Treaty.

Article XI

This Treaty, the English, Russian, French, Spanish and Chinese texts of which are equally authentic, shall be deposited in the archieves of the Depositary Governments. Duly certified copies of this Treaty shall be transmitted by the Depositary Governments to the Governments of the signatory and acceding States.

APPENDIX B

URANIUM ENRICHMENT CALCULATIONS

Natural uranium contains a mixture of isotopes, only one of which—U_{235}—is fissionable by the thermal (low energy) neutrons present in a light-water reactor. The concentration of this isotope in natural uranium is only 0.711 percent by weight and this concentration must be enhanced by enrichment to levels of 2 to 4 percent. This enhancement in conventional gaseous diffusion enrichment plants is a gradual process with many stages, consuming significant amounts of energy, and relies for separation on the fact that different isotopes have different masses. Because this process is thermodynamic in nature, the calculations of separative work are somewhat more complicated than for simple mass balances.

In the following we give the expressions that will allow the reader to calculate mass balances, enrichment requirements, and optimal tails assay.

Let c_f = concentration by weight of U_{235} in feed material (0.00711 for natural uranium);

       c_p = desired product enrichment level by weight (e.g., 0.030);

       c_t = tails assay desired (e.g., 0.002).

and let F, P, T be, respectively, the amounts of feed, product and tails materials (all isotopes, and quantities stated in terms of contained elemental uranium, independent of chemical form).

Then the mass balances for total uranium and for the contained U_{235} are:

$$F = P + T$$
$$F.c_f = P.c_p + T.c_t$$

and the ratio of feed-to-product and tails-to-product may be written as (with natural uranium feed and 0.20% tails assay as an example):

$$\frac{F}{P} = \frac{c_p - c_t}{c_f - c_t} = \frac{c_p - 0.002}{0.00711 - 0.002}$$

$$\frac{T}{P} = \frac{c_p - c_f}{c_f - c_t} = \frac{c_p - 0.00711}{0.00711 - 0.002}$$

The quality of enrichment, in kilograms separative work units (SWUs) needed to make one kilogram of final product at a given enrichment level is:

$$\frac{S}{P} = V_p + \frac{T}{P}.V_t - \frac{F}{P}V_f$$

where $\quad V_i = (2c_i - 1)ln\,(\frac{c_i}{1 - c_i}) \qquad i = p, t, f$

This formula is completely general and may be used for computing enrichment needs beginning with material that has previously been enriched (or depleted). We may also use this formula to derive an expression for the economically optimal tails assay, given the prices of enrichment (P_{swu}) and uranium feed (P_{uc}). The latter includes the cost of conversion to UF_6, the material actually used in the enrichment process, as well as the cost of the contained uranium. The total cost of a kilogram of fuel enriched to a given concentration c_p beginning with feed having a U_{235} concentration c_f is just:

$$C = P_{uc} \cdot \frac{F}{P} + P_{swu} \cdot \frac{S}{P}$$

We can now find the minimum of this total cost C as a function of the tails assay, c_t (by differentiating this expression with respect to c_t and setting the result equal to zero). Doing so, we find that the optimal tails assay is related to the ratio of the price of (converted) uranium feed to that of enrichment:

$$\frac{P_{cu}}{P_{swu}} = V_f - \frac{(2c_f - 1)}{(2c_{to} - 1)}\,V_t - \frac{(c_f - c_{to}).(2c_{to} - 1)}{c_{to}.(1 - c_{to})}$$

where c_{to} is the optimal tails assay.

This expression can be calculated for various values of c_{to}. The result is displayed graphically in Figure 1–6.

Table B-1
Historical World Uranium Production (MTU).

Year	U.S.	Canada	Niger	France[1]	Gabon	S. Africa	Namibia	Belgian Congo[2]	Australia	Others[3]	Total	Cumulative
1948	200	100						1,300			1,600	1,600
1949	365	100		75				1,400			1,940	3,540
1950	420	200		100				2,040			2,760	6,300
1951	690	200		100				1,600			2,590	8,890
1952	770	430		100		100		1,070			2,470	11,360
1953	1,500	800		100		780		970			4,150	15,510
1954	2,700	970		100		1,290		970	100		6,130	21,640
1955	3,100	970		180		2,153		970	180		7,553	29,193
1956	6,200	1,730		200		3,445		1,000	220		12,795	41,988
1957	7,567	5,075		369		4,383		1,200	321		18,915	60,903
1958	10,766	10,305		807		4,806		1,822	466		28,972	89,875
1959	13,381	12,227		896		4,960		1,784	859		34,107	123,982
1960	14,457	9,786		1,038		4,922		915	934		32,052	156,034
1961	14,226	7,382		1,450		4,206		123	1,197		28,584	184,618
1962	13,150	6,459		1,523		3,876			1,047		26,055	210,673
1963	11,304	6,459		1,584		3,460			917		23,724	234,397
1964	10,689	5,614		1,580		3,422			282		21,587	255,984
1965	8,151	3,384		1,611		2,261			260		15,667	271,661
1966	7,536	2,999		1,622		2,522			260		14,939	286,600
1967	8,228	2,845		1,692		2,476			260		15,501	302,101
1968	9,300	3,014		1,720		3,050			260		17,344	319,445
1969	8,900	3,430		1,180	500	3,080			254	250	17,594	337,039
1970	9,900	3,530		1,250	400	3,167			254	110	18,611	355,650
1971	9,470	3,830	430	1,250	540	3,220			0	194	18,934	374,584
1972	9,900	4,000	867	1,545	210	3,197			0	168	19,887	394,471
1973	10,200	3,710	948	1,616	402	2,735			0	162	19,773	414,244
1974	8,900	3,420	1,117	1,673	436	2,711			0	215	18,472	432,716
1975	8,900	3,510	1,306	1,742	800	2,488			0	334	19,080	451,796
1976	9,800	4,850	1,460	1,871	850	2,758	650		360	348	22,947	474,743
1977	11,500	5,790	1,609	2,097	907	3,360	2,340		356	390	28,349	503,092
1978	14,200	6,800	2,060	2,183	1,022	3,960	2,697		516	452	33,890	536,982
1979	14,400	6,820	3,620	2,362	1,100	4,797	3,840		705	465	38,109	575,091
1980	16,800	7,150	4,100	2,634	1,033	6,146	4,042		1,561	499	43,965	619,056
1981	14,793	7,720	4,360	2,553	1,022	6,131	3,971		2,860	486	43,892	662,948
1982	10,331	8,080	4,259	2,859	970	5,816	3,776		4,453	787	41,331	704,279

1. Before 1969, French production includes that from Gabon and other "affiliates."
2. The Belgian Congo is now Zaire.
3. Other includes: Argentina, Brazil, Japan, West Germany, Portugal, and Spain.

Sources: Data through 1956 are from U.S. Geological Survey Professional Paper 820, 1973. Data for 1957–1982 are from *Uranium Resources, Production and Demand*, OECD Nuclear Energy Agency and the International Atomic Energy Agency, Paris, 1970, 1973, 1976, 1977, 1979, 1982, 1983.

Table B-2
Historical Reserve and Resource Estimates (1000 MTU).

Country	1967		1970		1973		1975		1977		1979		1981	
	R.A.	E.A.	R.A.	E.A.	R.A.	E.A.	R.A.	E.A.	R.A.	E.A.	R.A.	E.A.	R.A.	E.A.
Algeria	—	—	—	—	—	—	28	—	28	50	28	0	26	0
	—	—	—	—	—	—	—	—	0	0	0	5	0	0
Argentina	7	16	8	17	9	14	9	15	18	24	23	9	25	4
	9	25	9	25	8	23	11	24	0	0	5	5	5	10
Australia	8	2	17	5	71	79	243	80	289	44	292	46	294	264
	2	1	7	5	30	29	—	—	7	5	8	8	23	21
Austria	—	—	—	—	—	—	—	—	2	0	2	0	0	1
	—	—	—	—	—	—	—	—	0	0	0	0	0	1
Bolivia	—	—	—	—	—	—	—	—	0	0	0	0	—	—
	—	—	—	—	—	—	—	—	0	1	0	1	—	—
Brazil	—	—	1	—	—	3	10	9	18	8	74	100	120	81
	—	—	—	—	1	—	1	—	0	1	0	0	0	0
Canada	154	100	178	100	185	122	144	22	167	15	215	19	230	28
	223	131	177	131	190	219	324	95	392	264	369	358	358	402
Central African Republic	—	—	8	—	8	8	8	8	8	8	18	8	18	0
	—	—	—	—	—	—	—	—	0	0	0	0	0	0
Chile	—	—	—	—	—	—	—	—	0	5	0	5	0	0
	—	—	—	—	—	—	—	—	0	0	0	0	0	7
Denmark (Greenland)	—	—	—	—	6	10	6	10	0	0	0	0	0	0
	4	—	4	—	—	—	—	—	6	9	27	16	27	16
Finland	—	—	—	—	1	—	2	—	1	0	0	0	0	0
	—	—	—	—	—	—	—	—	2	0	3	0	3	0

Country														
France	35	15	35	19	37	24	37	25	37	24	39	26	60	28
	4	8	7	12	20	25	18	15	15	20	16	20	16	18
Gabon	3	3	10	5	20	5	20	5	20	5	20	5	19	0
	—	—	—	5	—	5	—	5	0	5	0	5	2	10
Germany	—	—	—	—	—	—	1	1	2	3	4	3	1	2
	—	—	—	—	—	—	1	3	1	1	0	1	4	7
India	—	—	—	—	—	—	3	1	30	24	30	0	32	1
	2	1	2	1	2	1	26	23	0	0	0	24	0	24
Italy	1	—	1	—	1	—	—	—	1	1	0	1	0	0
	—	—	—	—	—	—	1	1	0	0	2	0	2	2
Japan	—	—	2	—	3	—	1	—	8	0	8	0	8	0
	3	—	4	—	4	—	7	—	0	0	0	0	0	0
Korea	—	—	—	—	—	—	—	—	0	0	0	0	0	—
	—	—	—	—	—	—	2	—	3	0	3	0	11	—
Madagascar	—	—	—	—	—	—	—	—	0	0	0	0	—	—
	—	—	—	—	—	—	—	—	0	2	0	2	—	—
Mexico	—	—	1	—	1	—	5	—	5	2	6	2	3	4
	—	—	1	—	1	—	1	—	0	0	0	0	0	3
Namibia	—	—	—	—	—	—	—	—	—	—	117	30	119	30
	—	—	—	—	—	—	—	—	—	—	16	23	16	23
Niger	9	10	20	29	40	20	40	20	160	53	162	53	160	53
	10	—	10	10	10	10	10	10	0	0	0	0	0	0

Table B–2. *(continued)*

Country	1967 R.A.	1967 E.A.	1970 R.A.	1970 E.A.	1973 R.A.	1973 E.A.	1975 R.A.	1975 E.A.	1977 R.A.	1977 E.A.	1979 R.A.	1979 E.A.	1981 R.A.	1981 E.A.
Philippines	—	—	—	—	—	—	—	—	0	0	0	0	—	—
Portugal	7	5	7	—	6	1	7	—	7	2	7	1	7	1
	—	9	6	12	6	10	—	—	2	0	2	1	2	1
Somalia	—	—	—	—	—	—	—	—	0	6	0	5	0	7
	—	—	—	—	—	—	—	—	0	3	0	2	0	3
South Africa	158	50	154	50	202	62	186	90	306	42	246	145	247	109
	12	27	12	27	8	26	6	68	34	38	54	85	84	91
Spain	9	3	9	8	9	8	10	94	7	0	10	0	12	4
	—	23	—	—	—	—	9	98	9	0	9	0	8	0
Sweden	269	38	269	—	270	40	—	300	1	300	1	299	0	38
	—	—	—	—	—	—	—	—	3	0	0	3	0	44
Turkey	—	—	—	—	2	1	7	1	4	0	2	2	2	2
	—	—	—	—	—	—	—	—	0	0	0	0	0	0
United Kingdom	—	—	—	—	—	—	2	4	0	0	0	8	0	0
	—	—	—	—	—	—	—	—	0	7	0	0	0	7
United States	250	154	390	231	538	231	500	312	838	215	777	381	681	416
	139	77	192	108	259	141	320	134	523	120	531	177	362	243
Yugoslavia	—	—	—	—	6	—	4	2	5	2	5	2	—	—
	—	—	—	—	10	—	—	15	5	16	5	16	—	—
Zaire	5	—	—	—	2	—	2	2	2	0	2	0	2	0
	—	—	—	—	—	—	—	—	2	0	2	0	2	0

Totals													
535	430	643	669	867	917	1,085	1,005	1,649	1,511	1,842	1,506	1,747	1,603
529	417	579	459	682	619	731	683	545	586	730	964	541	1,106
1,064	957	1,222	1,128	1,549	1,536	1,816	1,688	2,194	2,097	2,572	2,470	2,288	2,709

Note: The first row of entries for each country is for the lower cost category of reserves and resources while the second is for the higher. See text discussion, Chapter 4.

Sources:

OECD Nuclear Energy Agency and the International Atomic Energy Agency, *Uranium Resources*, Paris, 1967.

OECD Nuclear Energy Agency and the International Atomic Energy Agency, *Uranium Production and Short-Term Demand*, Paris, 1969.

OECD Nuclear Energy Agency and the International Atomic Energy Agency, *Uranium Resources, Production and Demand*, Paris, 1970, 1973, 1976, 1977, 1979, 1982.

Table B–3

Utility Nuclear Growth Estimates (*GWE-All Reactor Types*)
(Non-U.S. WOCA)

Capacity Projected for Year	Nuclear News Surveys of					
	6/75	*6/77*	*6/79*	*12/80*	*12/81*	*12/82*
1975	33.9					
1976	46.9					
1977	61.2	48.2				
1978	78.5	61.8				
1979	93.9	79.7	64.7			
1980	111.2	98.8	78.7	64.2		
1981	123.2	115.0	99.1	83.5	75.3	
1982	129.9	134.6	114.9	99.0	93.7	85.0
1983	130.8	150.1	135.2	114.6	119.9	105.0
1984	130.8	157.1	146.8	135.4	132.2	126.3
1985	130.8	161.6	158.9	152.8	149.5	148.3
1986	130.8	162.3	163.7	164.5	163.0	160.7
1987	130.8	162.3	165.9	170.5	174.6	172.6
1988	130.8	162.3	167.7	179.3	185.9	182.1
1989	130.8	162.3	168.6	185.7	189.7	191.6
1990	130.8	162.3	170.5	189.2	190.6	196.1
1995	130.8	162.3	172.1	190.2	191.7	202.8

Source: Based on *Nuclear News* (American Nuclear Society) surveys. August 1975, August 1977, August 1979, February 1981, February 1982, and February 1983. Only reactors in operation, under construction or on order with a startup date specified are included.

Table B-4

Present Plans Reactor Capacity Growth Scenario¹ (GWe)

	1981	1982	1983	1984	1985	1986	1987	1988	1989	1990	1995	2000
Belgium	1.6	2.6	3.4	5.4	5.4	5.4	5.4	5.4	5.4	5.4	5.4	5.4
Canada	5.5	5.5	7.3	9.8	10.3	11.1	11.8	12.7	13.6	13.6	15.3	15.3
Finland	1.5	2.2	2.2	2.2	2.2	2.2	2.2	2.2	2.2	2.2	2.2	2.2
France	19.0	22.7	29.3	34.4	43.1	46.5	51.2	53.7	56.3	56.3	56.3	56.3
Germany, FR	8.6	9.8	11.1	13.9	17.6	20.0	20.0	21.3	23.7	26.2	28.8	28.8
Greece	—	—	—	—	—	—	—	—	—	—	—	—
Italy	1.3	1.3	1.3	1.3	1.3	2.3	3.3	3.3	3.3	3.3	3.3	3.3
Japan	15.0	16.6	16.6	19.1	22.8	24.8	24.8	24.8	25.4	25.9	26.2	26.2
Netherlands	.5	.5	.5	.5	.5	.5	.5	.5	.5	.5	.5	.5
Spain	2.0	2.0	4.7	6.5	6.5	8.4	10.4	11.4	12.4	12.4	12.4	12.4
Sweden	6.4	6.4	7.3	7.3	9.4	9.4	9.4	9.4	9.4	9.4	9.4	9.4
Switzerland	1.9	1.9	1.9	2.9	2.9	2.9	2.9	2.9	2.9	2.9	3.8	3.8
Turkey	—	—	—	—	—	—	—	—	—	—	—	—
United Kingdom	8.0	8.6	11.7	11.7	11.7	11.7	13.1	14.4	14.4	14.4	14.4	14.4
United States	56.9	59.2	72.5	86.3	96.4	107.4	115.4	118.8	120.1	121.3	123.9	126.1
Developing Nations²	4.0	4.9	7.6	11.2	14.5	15.4	17.5	20.0	22.0	23.5	24.7	24.7
Total	132.2	144.2	177.4	212.5	244.6	268.0	287.9	300.8	311.6	317.3	326.6	328.8

Notes:

1. From *Nuclear News*, February 1983 (Utility survey conducted as of December 31, 1982). Data were not included for Eastern Europe, USSR nor China.

2. Developing Nations consists of Argentina, Brazil, Egypt, India, South Korea, Mexico, Pakistan, Philippines, South Africa, Taiwan, and Thailand.

Table B-5
Moderate Reactor Capacity Growth Scenario[1] (*GWe*)

	1981	1982	1983	1984	1985	1986	1987	1988	1989	1990	1995	2000
Belgium	1.7	3.5	3.5	3.5	5.4	5.4	5.4	5.4	5.4	5.4	6.4	7.4
Canada	5.2	5.9	7.6	9.3	10.1	10.9	11.6	12.5	13.4	14.3	18.7	23.0
Denmark	—	—	—	—	—	—	—	—	—	—	—	—
Finland	2.2	2.2	2.2	2.2	2.2	2.2	2.2	2.2	2.2	2.2	3.2	4.2
France	21.8	24.5	26.3	32.7	35.6	39.9	45.1	49.0	52.3	54.8	67.2	77.8
Germany, FR	6.4	8.2	10.0	10.0	10.0	10.0	10.0	10.0	10.3	10.6	14.8	18.1
Greece	—	—	—	—	—	—	—	—	—	—	.9	1.8
Italy	1.3	1.3	1.3	1.3	1.3	1.3	2.2	3.2	3.2	4.2	8.8	12.8
Japan	15.0	17.0	17.0	19.0	21.0	25.0	26.0	30.0	34.0	37.0	53.0	68.0
Netherlands	.5	.5	.5	.5	.5	.5	.5	.5	.5	.5	.5	.5
Portugal	—	—	—	—	—	—	—	—	—	—	—	—
Spain	2.0	3.9	4.8	6.7	7.6	8.6	9.5	10.5	11.5	12.5	16.5	21.3
Sweden	6.4	7.3	7.3	7.3	8.4	9.4	9.4	9.4	9.4	9.4	9.4	9.4
Switzerland	1.9	1.9	1.9	2.9	2.9	2.9	2.9	2.9	2.9	3.8	4.8	5.8
Turkey	—	—	—	—	—	—	—	—	—	—	1.2	1.8
United Kingdom	6.4	8.2	10.0	10.0	10.0	10.0	10.0	10.0	10.3	10.6	14.8	18.1
United States	56.0	63.0	74.0	85.0	91.0	99.0	105.0	111.0	115.0	117.0	131.0	150.0
Developing Nations[2]	6.2	8.6	11.6	13.6	20.0	23.9	26.2	29.4	34.6	38.7	62.8	92.0
Total	133.0	156.0	178.0	204.0	226.0	249.0	266.0	286.0	305.0	321.0	414.0	512.0

Notes:

1. Based on NUKEM Market Report of May 1982. Data were not included for Eastern Europe, USSR nor China.

2. Developing Nations consists of Argentina, Brazil, Egypt, India, South Korea, Mexico, Pakistan, Philippines, South Africa, Taiwan, and Thailand.

INDEX

ABOUT THE AUTHOR

THOMAS L. NEFF is Director of the International Energy Studies Program at the Energy Laboratory of the Massachusetts Institute of Technology. His research interests include economic and political dimensions of international energy markets, natural resource development, and national and international security issues. He has published widely on these topics and has served as an advisor to numerous government agencies and private organizations in the United States and abroad. Prior to his present position, he served as chief staff officer of the Nuclear Energy Policy Study that presented its recommendations to the president in 1977. Dr. Neff received his Ph.D. in physics from Stanford University in 1972.